Sustainable Marine Food and Feed Production Technologies

This valuable reference book covers the application of marine resources like algae, fishes, and shrimp to address food challenges of the world. It compiles technological advancements in employing these resources for food and food supplements to enhance human health. The book includes chapters from international experts. The book discusses interesting topics like exploitation of marine wastes as nutraceuticals, cultivation, processing, and production of seafood and includes a section on other applications of marine organisms, such as the removal of pollutants from wastewater. This book is meant for graduate students and researchers in food science. It is also useful to experts in the food industry.

Sustainable Marine Food and Feed Production Technologies

Edited By

Anil Kumar Patel
Institute of Aquatic Science and Technology, National Kaohsiung University of Science and Technology, Kaohsiung, Taiwan & Centre for Energy and Environmental Sustainability, Lucknow, India

Reeta Rani Singhania
Institute of Aquatic Science and Technology, National Kaohsiung University of Science and Technology, Kaohsiung, Taiwan & Centre for Energy and Environmental Sustainability, Lucknow, India

Cheng-Di Dong
Institute of Aquatic Science and Technology, National Kaohsiung University of Science and Technology, Kaohsiung, Taiwan

Ashok Pandey
Centre for Innovation and Translational Research, CSIR-Indian Institute of Toxicology Research, Lucknow, India & Sustainability Cluster, School of Engineering, University of Petroleum and Energy Studies, Dehradun, India & Centre for Energy and Environmental Sustainability, Lucknow, India

CRC Press is an imprint of the
Taylor & Francis Group, an **informa** business

First edition published 2023
by CRC Press
2385 Executive Center Drive, Suite 320, Boca Raton, FL 33431

and by CRC Press
4 Park Square, Milton Park, Abingdon, Oxon, OX14 4RN

CRC Press is an imprint of Taylor & Francis Group, LLC

© 2024 selection and editorial matter, Anil Kumar Patel, Reeta Rani Singhania, Cheng-Di Dong, Ashok Pandey; individual chapters, the contributors

Reasonable efforts have been made to publish reliable data and information, but the author and publisher cannot assume responsibility for the validity of all materials or the consequences of their use. The authors and publishers have attempted to trace the copyright holders of all material reproduced in this publication and apologize to copyright holders if permission to publish in this form has not been obtained. If any copyright material has not been acknowledged please write and let us know so we may rectify in any future reprint.

Except as permitted under U.S. Copyright Law, no part of this book may be reprinted, reproduced, transmitted, or utilized in any form by any electronic, mechanical, or other means, now known or hereafter invented, including photocopying, microfilming, and recording, or in any information storage or retrieval system, without written permission from the publishers.

For permission to photocopy or use material electronically from this work, access www.copyright.com or contact the Copyright Clearance Center, Inc. (CCC), 222 Rosewood Drive, Danvers, MA 01923, 978–750–8400. For works that are not available on CCC please contact mpkbookspermissions@tandf.co.uk

Trademark notice: Product or corporate names may be trademarks or registered trademarks and are used only for identification and explanation without intent to infringe.

ISBN: 978-1-032-35448-4 (hbk)
ISBN: 978-1-032-35449-1 (pbk)
ISBN: 978-1-003-32694-6 (ebk)

DOI: 10.1201/9781003326946

Typeset in Times
by Apex CoVantage, LLC

Contents

Preface vii
Editors ix
Contributors xi

Chapter 1 Marine Algae as a Source of Carotenoids: Bioactivity and Health Benefits 1

Anil Kumar Patel, Vaibhav Sunil Tambat, Akash Pralhad Vadrale, Prashant Kumar, Reeta Rani Singhania, Chiu-Wen Chen, Ashok Pandey and Cheng-Di Dong

Chapter 2 Innovative Extraction Methods to Obtain Bioactive Compounds from Aquatic Biomass 21

Ana Lucía Sarmiento-Padilla, Blanca E. Morales-Contrerasa, Rohit Saxena, Ruth Belmares, Adriana M. Bonilla Loaiza, K.D. González-Gloria, Araceli Loredo-Treviño, Rosa M. Rodríguez-Jasso and Héctor A. Ruiz

Chapter 3 Bioactive Polysaccharides from Macroalgae 31

Latifa Tounsi, Faiez Hentati, Olga Babich, Stanislav Sukhikh, Roya Abka Khajouei, Anil Kumar Patel, Reeta Rani Singhania, Imen Fendri, Slim Abdelkafi and Philippe Michaud

Chapter 4 Marine Microalgae in Food and Health Applications 45

Tirath Raj, K Chandrasekhar, Raj Morya and Sang-Hyoun Kim

Chapter 5 Filler Feed from Marine Micro- and Macroalgae 61

Dillirani Nagarajan, Duu-Jong Lee and Jo-Shu Chang

Chapter 6 Marine Collagen: Valorization of Marine Wastes for Health Care and Biomaterials 77

Grace Sathyanesan Anisha, Anil Kumar Patel and Reeta Rani Singhania

Chapter 7 Application of Functional Aquafeed in Sustainable Aquaculture 89

Yu-Hung Lin, Winton Cheng and Hsin-Wei Kuo

Chapter 8 Sustainable Development for Shrimp Culture: A Critical Analysis 103

Kaushik Dey and Sanghamitra Sanyal

Chapter 9 Biochar as a Potential Candidate for Removal of Pollutants from Aquaculture 121

Thanh Binh Nguyen, Xuan-Thanh Bui, Quoc-Minh Truong, Thi-Kim-Tuyen Nguyen, Thi-Bao-Chau Ho, Phung-Ngoc-Thao Ho, Van-Re Le, Van-Anh Thai, Hien-Thi-Thanh Ho, Van Dien Dang and Cheng-Di Dong

Chapter 10 Valorization of Seafood Processing Discards 131

Ya-Ting Chen and Shu-Ling Hsieh

Chapter 11 Effect of Microplastics on Marine Ecosystems 145

De-Sing Ding and Anil Kumar Patel

Chapter 12 Role of Marine Algae for GHG Reduction/CO2 Sequestration: Strategies and Applications 157

Xinwei Sun, Andrei Mikhailovich Dregulo, Su Zhenni, Obulisamy Parthiba Karthikeyan, Sharareh Harirchi, El-Sayed Salama, Yue Li, Raveendran Sindhu, Parameswaran Binod, Ashok Pandey and Mukesh Kumar Awasthi

Index 167

Preface

The book entitled *Sustainable Marine Food and Feed Production Technologies* is published by CRC Press in partnership with Taylor & Francis Group and the Biotech Research Society, India (BRSI).

"FOOD CAN BE DESIGNED SO THAT NATURE THRIVES RATHER THAN BENT TO PRODUCE FOOD"

Globally the increasing population and concomitant increased need to fulfill food obligations has put pressure on nature. It necessitates utilization of available food resources in the best possible way. Marine food resources can be instrumental in resolving these challenges, and this book explores the application of marine algae, shrimps, and fishes. The technological advancements in the extraction of the nutritional components from marine resources enable the efficient utilization of these resources for nutraceutical application.

This book includes chapters from renowned scientists and researchers in their field who have explicitly transferred their accumulated experience and knowledge in the form of chapters that can ignite young minds to march forward to take society towards a contented future. The book's first chapter elucidates the importance of carotenoids from marine algae for human health. The authors explain the kind of bioactivities the pigments possess, such as antioxidant, antidiabetic, anticancer, neuroprotective, and immunoprophylactic, besides their general coloring properties. Relevant advancements in research directions and the commercial importance of the pigments are presented. The second chapter, "Innovative Extraction Methods to Obtain Bioactive Compounds from Aquatic Biomass," gives an account of various available techniques for the extraction of bioactive compounds from aquatic biomass and the challenges of those techniques, which allows careful selection of extraction technique to fulfil the needs of the process and the products. Chapter 3 deals with marine macroalgal polysaccharides from red, green, and brown macroalgal biomass such as fucoidan and ulvan, which possess immense health benefits. Advancements and challenges in the field are presented with a strong message for their utilization in nutraceuticals, looking at its immense bioactivity. Different processes accomplish the refinement and separation of bioactive substances. Several substances derived from marine microalgae have revealed promise as antioxidants and therapies for high blood pressure, among other health issues. Chapter 4 deals with food and medicinal applications of marine microalgae, which can be easily dealt with like other microorganisms.

Chapter 5 is on filler feed from marine algae. This chapter discusses the role of microalgae and macroalgae biomass as filler feed in aquaculture. Biomass constituents of algae that confer potential health benefits are discussed, and various studies representing the use of algal biomass as aquaculture feeds are summarized. Since sustainability and maintenance of ecological balance are critical for the fishing industry, various alternative aquaculture feeds are being investigated for the replacement of fishmeal and oil. The next chapter explores marine collagen for health and biomaterial applications. Marine collagen obtained from marine invertebrates and fishes has several advantages, such as low immunogenicity, high mechanical adaptability, biocompatibility, and biodegradability. Marine collagen and its derived peptides have several beneficial biological actions, such as antioxidant, antimicrobial, anti-obesogenic, and anti-diabetic actions. Marine collagen also finds applications in food, nutraceuticals, and cosmetics. Commercial-scale production of marine collagen offers the possibility of biotechnological valorization of refuse from the marine food industry. The success of aquaculture is largely dependent on aquafeed quality. The protein requirement for aquatic animals is commonly high, and dietary protein requirement is mainly fulfilled from marine-derived proteins, such as fish meal. The chapter reviews the application of functional aquafeed in sustainable aquaculture. The roles of functional feed additives in plant-based feed and strategies to replace the use of antibiotics are documented. Chapter 8 is on sustainable development for shrimp culture, and the authors critically analyze ongoing practice and discuss advancement with a case study.

The next chapters address the issue of removal of pollutants from aquaculture waste. Chapter 9 presents biochar as a potent candidate for removal of pollutants from aquaculture waste. The chapter assesses recent advances and research in engineered biochar, with a particular emphasis on water remediation, to contribute innovative ideas for future research involving engineered biochar. Chapter 10 deals with seafood processing discards for extracting components of nutritional value, which is an innovative idea. This chapter not only presents the sources of by-products and their impact on the environment, but also examines the enhanced research and application value of by-products using biological and physical methods. The objective of this chapter is to understand the sustainable value of seafood processing by-products and increase people's awareness for sustainable marine environmental development. Then Chapter 11 presents a very sensitive issue: the effect of microplastics on marine ecosystems. It presents the harm and impact of microplastics on marine life and also discuss the economic threats to the aquaculture industry. The final chapter investigates the role of marine algae in greenhouse gas reduction. In this chapter, the biological characteristics of marine algae and principles and challenges are presented; the carbon sequestration factors of marine algae are summarized, and specific ways to improve the carbon sequestration efficiency of marine algae are proposed; subsequently, the physiological mechanisms of carbon sequestration in marine microalgae (especially carbon-concentrating mechanisms) are presented and recent advances are discussed, and the limitations of carbon sequestration in marine microalgae are presented.

Thus, the chapters from international experts discusses interesting topics like the exploitation of marine waste as nutraceuticals, cultivation, processing, and production of seafood. The book also includes a section on other applications of marine organisms, such as the removal of pollutants from wastewater. This book is meant for graduate students and researchers in food science. It is also useful to experts in the food industry.

We sincerely appreciate the contribution of scientific experts from different countries for sharing their knowledge on marine food and feed technologies. We strongly believe that this book provides enriched scientific information that will be useful for researchers, students, academicians, and industry experts in marine food technology. We are grateful to the reviewers for their sincere efforts and acknowledge them for their contribution in the critical review of the chapters. We sincerely acknowledge the BRSI for providing us the opportunity to prepare this book and guiding us during the editing and publication process to bring this book into its final form. We thank the team of Taylor & Francis Group, comprising Dr. Gagandeep Singh, Senior Publisher; Ms. Madhurima Kahali, Editor II (Life Sciences); Ms Neha Bhatt, Editorial Assistant; CRS Press; and the entire team of the CRC Press–Taylor & Francis Group for their consistent support during the publication process of this book.

Editors
Anil Kumar Patel
Reeta Rani Singhania
Cheng-Di Dong
Ashok Pandey

Editors

Dr. Anil Kumar Patel is currently an associate professor at the Institute of Aquatic Science and Technology, National Kaohsiung University of Science and Technology, Kaohsiung, Taiwan. He is also Senior Scientist (Honorary) at the Centre for Energy and Environmental Sustainability—India. He obtained his MSc degree in biotechnology from Guru Ghasidas University, Bilaspur, India, and PhD in biotechnology from the North Maharashtra University, Jalgaon, India. His broad post-doctoral research was carried out in several overseas labs basically on waste-to-wealth. The area of his current work is mainly environmental bioengineering, broadly focused on waste-to-fuel/non-fuel product programs utilizing bacterial and algal systems. Previously he worked as research professor at Department of Chemical and Biological Engineering, Korea University, Seoul, and HKBU, Hong Kong, on algal biorefinery projects and solid waste management. He worked at University of Hawaii, USA, on the removal of sulfur from the gaseous phase of treated S-containing wastewaters by applying designer biochar, which was used for sulfur nutrition in plants. He gained experience in biohythane production using ethanol plant wastes at DBT-IOC Bioenergy Centre, India, and on membrane bioreactors for continuous biohydrogen production as well as VFA separation from industrial waste at Institute Pascal, France. He has been a visiting researcher at EPFL, Switzerland, to work on biomass pretreatments. In these studies, both bench-scale and pilot-scale studies were undertaken with techno-economic analysis. He has over 180 publications/communications, which include 4 patents, 150 original and review articles, 22 book chapters, and 2 conference proceedings.

Dr. Reeta Rani Singhania is currently a research associate professor at the Department of Sustainable Environment Research Center, National Kaohsiung University of Science and Technology (NKUST), Kaohsiung, Taiwan. She is also Senior Scientist (Honorary) at the Centre for Energy and Environmental Sustainability India. She worked at CSIR-National Institute for Interdisciplinary Science and Technology, Trivandrum, for PhD studies and obtained a degree in biotechnology in 2011. During 2011–2012, she worked as a post-doctoral fellow at University Blaise Pascal, Clermont-Ferrand, France, and 2012–2017 at DBT-IOC Centre for Advanced Bioenergy Research, Faridabad, as DBT-Bioscience Energy Overseas Fellow. Subsequently, she joined the Centre for Energy and Environmental Sustainability, India, and worked as chief scientist during 2018–2020. Her major research interests are in microbial and enzyme technology and bioprocess technology, with a current focus on biofuels from lignocellulosic biomass. She has 16 patents and has edited three books and published more than 170 research articles, book chapters, and conference communications, with more than 9700 citations and an h index of 44 (Google Scholar).

Professor Cheng-Di Dong is a distinguished professor at the Department of Marine Environmental Engineering at National Kaohsiung University of Science and Technology (NKUST). He is also the dean of the College of Hydrosphere Science and the director of Sustainable Environment Research Center (SERC) of NKUST. He earned a PhD in environmental engineering from the University of Delaware in 1993. His current research interests are broadly clustered in the areas of monitoring and remediation of contaminated water, soil, and sediments. He is particularly interested in developing the fate and transport of contaminants in marine sediments and remediation of contaminated sediments for the purpose of reuse sediment materials using physical, chemical, and biological processes. Prof. Dong has participated in the scientific committee of several conferences and associations and serves as a reviewer for a wide range of international renowned journals. He has published more than 400 papers in leading international journals, 8 patents, and 5 book chapters and edited 8 special issues of scientific journals.

Professor Ashok Pandey is currently distinguished scientist at the Centre for Innovation and Translational Research, CSIR-Indian Institute of Toxicology Research, Lucknow, India, and distinguished professor at the Sustainability Cluster, School of Engineering, University of Petroleum and Energy Studies, Dehradun, India. He is also Executive Director (Honorary) at the Centre for Energy and Environmental Sustainability India. Formerly, he was Eminent Scientist at the Center of Innovative and Applied Bioprocessing, Mohali and chief scientist and head of Biotechnology Division and Centre for Biofuels at CSIR's National Institute for Interdisciplinary Science and Technology, Trivandrum. His major research and technological development interests are industrial and environmental biotechnology and energy biosciences, focusing on biomass to biofuels and chemicals, waste to wealth and energy, industrial enzymes, and so on. He has 16 patents, 120 books, ~950 papers and book chapters, with an h index of 127 and more than 70,000 citations (Google Scholar). Professor Pandey is the recipient of many national and international awards and honors, which include highly cited researcher (Clarivate Analytics, Web of Science and Stanford University reports). distinguished fellow, the Biotech Research Society, India (2021) and fellow of several national and international academies, including The World Academy of Sciences, etc.

Contributors

Slim Abdelkafi
Laboratoire de Genie Enzymatique et Microbiologie, Equipe de Biotechnologie des Algues, Ecole Nationale d'Ingenieurs de Sfax
Universite de Sfax
Sfax, Tunisie

Grace Sathyanesan Anisha
Post-graduate and Research Department of Zoology
Government College for Women
Thiruvananthapuram, Kerala, India

Mukesh Kumar Awasthi
College of Natural Resources and Environment
Northwest A&F University
Yangling, Shaanxi Province, China

Olga Babich
Institute of Living Systems, Immanuel Kant Baltic Federal University
Kaliningrad, Russia
and
Belmares Ruth Biorefinery Group, Food Research Department, School of Chemistry Autonomous University of Coahuila
Coahuila, Mexico

Ruth Belmares
Biorefinery Group, Food Research Department, School of Chemistry Autonomous University of Coahuila
Coahuila, Mexico

Parameswaran Binod
Microbial Processes and Technology Division
CSIR-National Institute for Interdisciplinary Science and Technology
Thiruvananthapuram, India

Xuan-Thanh Bui
Key Laboratory of Advanced Waste Treatment Technology & Faculty of Environment and Natural Resources
and
University of Technology, Vietnam National University Ho Chi Minh
Ho Chi Minh City, Vietnam

K. Chandrasekhar
Department of Biotechnology, Vignan's Foundation for Science, Technology and Research, Vadlamudi, Guntur
Andhra Pradesh, India

Jo-Shu Chang
Department of Chemical Engineering, National Cheng Kung University
Tainan, Taiwan
and
Department of Chemical and Materials Engineering
Tunghai University
Taichung, Taiwan
and
Research Center for Smart Sustainable Circular Economy,
Tunghai University
Taichung, Taiwan
and
Department of Chemical Engineering and Materials Science
Yuan Ze University
Chung-Li, Taiwan

Chiu-Wen Chen
Institute of Aquatic Science and Technology
and
Department of Marine Environmental Engineering
and
Sustainable Environment Research Center
National Kaohsiung University of Science and Technology
Kaohsiung City, Taiwan

Ya-Ting Chen
Department of Seafood Science National Kaohsiung University of Science and Technology
Kaohsiung, Taiwan

Winton Cheng
Department of Aquaculture National Pingtung
University of Science and Technology
Neipu, Pingtung, Taiwan

Van Dien Dang
Faculty of Biology-Environment, Ho Chi Minh
University of Food Industry
Ho Chi Minh City, Vietnam

Kaushik Dey
Government of West Bengal
West Bengal, India

De-Sing Ding
Institute of Aquatic Science and Technology
College of Hydrosphere Science
and
Department and Graduate Institute of Aquaculture
National Kaohsiung University of Science
  and Technology
Kaohsiung, Taiwan

Cheng-Di Dong
Institute of Aquatic Science and Technology
National Kaohsiung University of Science
  and Technology
and
Department of Marine Environmental
  Engineering
National Kaohsiung University of Science
  and Technology
and
Sustainable Environment Research Center
National Kaohsiung University of Science
  and Technology
Kaohsiung City, Taiwan

Andrei Mikhailovich Dregulo
Federal State Budgetary Educational Institution
  of Higher Education
Saint-Petersburg State University
Saint Petersburg, Russia

Imen Fendri
Laboratoire de Biotechnologie des Plantes
  Appliquee a l'Amelioration des Cultures,
Faculte des Sciences de Sfax
Universite de Sfax
Sfax, Tunisie

K.D. González-Gloria
Biorefinery Group, Food Research Department,
  School of Chemistry
Autonomous University of Coahuila
Saltillo, Coahuila, Mexico

Sharareh Harirchi
Swedish Centre for Resource Recovery
University of Boras
Boras, Sweden

Faiez Hentati
INRAE, URAFPA
Universite de Lorraine
Nancy, France

Hien-Thi-Thanh Ho
Faculty of Environment, School of Engineering
  and Technology
Van Lang University
Ho Chi Minh City, Vietnam

Phung-Ngoc-Thao Ho
Institute of Aquatic Science and Technology
National Kaohsiung University of Science
  and Technology
and
Department of Marine Environmental
  Engineering
National Kaohsiung University of Science
  and Technology
Kaohsiung City, Taiwan

Thi-Bao-Chau Ho
Institute of Aquatic Science and Technology,
National Kaohsiung University of Science
  and Technology
and
Department of Marine Environmental
  Engineering
National Kaohsiung University of Science
  and Technology
Kaohsiung City, Taiwan

Shu-Ling Hsieh
Department of Seafood Science
National Kaohsiung University of Science
  and Technology
Kaohsiung, Taiwan

Obulisamy Parthiba Karthikeyan
Sherbrooke Research and Development Center
Agriculture and Agri-Food Canada
Sherbrooke, Québec, Canada

Contributors

Roya Abka Khajouei
Department of Food Science and Technology, College of Agriculture
Isfahan University of Technology
Is-fahan, Iran

Sang-Hyoun Kim
School of Civil and Environmental Engineering,
Yonsei University
Seoul, Republic of Korea

Prashant Kumar
Institute of Aquatic Science and Technology
National Kaohsiung University of Science and Technology
Kaohsiung City, Taiwan

Hsin-Wei Kuo
Department of Aquaculture National Pingtung University of Science and Technology
and
General Research Service Center
National Pingtung University of Science and Technology
Neipu, Pingtung, Taiwan

Van-Re Le
Institute of Aquatic Science and Technology
National Kaohsiung University of Science and Technology
and
Department of Marine Environmental Engineering
National Kaohsiung University of Science and Technology
Kaohsiung City, Taiwan

Duu-Jong Lee
Department of Chemical Engineering
National Taiwan University
Taipei, Taiwan
and
Department of Mechanical Engineering
City University of Hong Kong
Kowloon Tang, Hong Kong

Yue Li
College of Natural Resources and Environment
Northwest A&F University
Yangling, Shaanxi Province, China

Yu-Hung Lin
Department of Aquaculture National Pingtung University of Science and Technology
Neipu, Pingtung, Taiwan

Adriana M. Bonilla Loaiza
Biorefinery Group, Food Research Department, School of Chemistry
Autonomous University of Coahuila
Saltillo, Coahuila, Mexico

Araceli Loredo-Treviño
Biorefinery Group, Food Research Department, School of Chemistry
Autonomous University of Coahuila
Saltillo, Coahuila, Mexico

Philippe Michaud
CNRS, SIGMA Clermont, Institut Pascal
Universite Clermont Auvergne
Clermont-Ferrand, France

Blanca E. Morales-Contrerasa
Biorefinery Group, Food Research Department, School of Chemistry
Autonomous University of Coahuila
Saltillo, Coahuila, Mexico
and
Departamento de Ingenieria Quimica y Bioquimica
Tecnologico Nacional de Mexico/I.T.
Durango, Durango, Mexico.

Raj Morya
School of Civil and Environmental Engineering
Yonsei University
Seoul, Republic of Korea

Dillirani Nagarajan
Department of Chemical Engineering
National Taiwan University
Taipei, Taiwan
and
Department of Chemical Engineering
National Cheng Kung University
Tainan, Taiwan

Thanh Binh Nguyen
Institute of Aquatic Science and Technology
National Kaohsiung University of Science and Technology
Kaohsiung City, Taiwan

Thi-Kim-Tuyen Nguyen
Institute of Aquatic Science and Technology
National Kaohsiung University of Science and Technology
and
Department of Marine Environmental Engineering
National Kaohsiung University of Science and Technology
Kaohsiung City, Taiwan

Ashok Pandey
Centre for Innovation and Translational Research
CSIR-Indian Institute of Toxicology Research
Lucknow
and
Sustainability Cluster, School of Engineering
University of Petroleum and Energy Studies
Dehradun
and
Centre for Energy and Environmental Sustainability
Lucknow, India

Anil Kumar Patel
Institute of Aquatic Science and Technology
National Kaohsiung University of Science and Technology
and
Department of Marine Environmental Engineering
National Kaohsiung University of Science and Technology
and
Sustainable Environment Research Center
National Kaohsiung University of Science and Technology
Kaohsiung City, Taiwan
and
Centre for Energy and Environmental Sustainability
Lucknow, India

Tirath Raj
School of Civil and Environmental Engineering
Yonsei University
Seoul, Republic of Korea

Sindhu Raveendran
Department of Food Technology
T K M Institute of Technology
Kollam, Kerala, India

Rosa M. Rodríguez-Jasso
Biorefinery Group, Food Research Department
School of Chemistry, Autonomous University of Coahuila
Saltillo, Coahuila, Mexico

Héctor A. Ruiz
Biorefinery Group, Food Research Department, School of Chemistry
Autonomous University of Coahuila
Saltillo, Coahuila, Mexico

El-Sayed Salama
Department of Occupational and Environmental Health, School of Public Health
Lanzhou University
Lanzhou City, Gansu Province, PR China

Sanghamitra Sanyal
SACT-1, Durgapur Government College
Durgapur, West Bengal, India

Ana Lucía Sarmiento-Padilla
Biorefinery Group, Food Research Department, School of Chemistry
Autonomous University of Coahuila
Saltillo, Coahuila, Mexico

Rohit Saxena
Biorefinery Group, Food Research Department, School of Chemistry
Autonomous University of Coahuila
Saltillo, Coahuila, Mexico

Reeta Rani Singhania
Department of Marine Environmental Engineering
National Kaohsiung University of Science and Technology
and
Sustainable Environment Research Center
National Kaohsiung University of Science and Technology
Kaohsiung City, Taiwan
and
Centre for Energy and Environmental Sustainability
Lucknow, India

Stanislav Sukhikh
Institute of Living Systems
Immanuel Kant Baltic Federal University
Kaliningrad

Contributors

Xinwei Sun
College of Natural Resources and Environment
Northwest A&F University
Yangling, Shaanxi Province, China

Vaibhav Sunil Tambat
Institute of Aquatic Science and Technology
National Kaohsiung University of Science
  and Technology
Kaohsiung City, Taiwan

Van-Anh Thai
Institute of Aquatic Science and Technology
National Kaohsiung University of Science
  and Technology
and
Department of Marine Environmental
  Engineering
National Kaohsiung University of Science
  and Technology
Kaohsiung City, Taiwan

Latifa Tounsi
CNRS, SIGMA Clermont, Institut Pascal
Universite Clermont Auvergne
Clermont-Ferrand, France
and
Laboratoire de Genie Enzymatique et
  Microbiologie
Equipe de Biotechnologie des Algues,
  Ecole Nationale d'Ingenieurs de Sfax,
  Universite de Sfax
Sfax, Tunisie

Quoc-Minh Truong
Institute of Aquatic Science and Technology
National Kaohsiung University of Science
  and Technology
and
Department of Marine Environmental
  Engineering
National Kaohsiung University of Science
  and Technology
Kaohsiung City, Taiwan

Akash Pralhad Vadrale
Institute of Aquatic Science and Technology
National Kaohsiung University of Science
  and Technology
Kaohsiung City, Taiwan

Zhenni Su
Department of Occupational and Environmental
  Health, School of Public Health
Lanzhou University
Lanzhou City, Gansu Province, China

1 Marine Algae as a Source of Carotenoids
Bioactivity and Health Benefits

Anil Kumar Patel, Vaibhav Sunil Tambat, Akash Pralhad Vadrale, Prashant Kumar, Reeta Rani Singhania, Chiu-Wen Chen, Ashok Pandey and Cheng-Di Dong

CONTENTS

1.1 Introduction 1
1.2 De Novo Synthesis of Algal Pigments 3
1.3 Important Carotenoid Structure, Classification, and Significance 4
    1.3.1 Astaxanthin 4
        1.3.1.1 Microalgal Production and Significance of Astaxanthin 4
        1.3.1.2 Bioactivity of Astaxanthin 5
    1.3.2 Lutein 7
        1.3.2.1 Microalgal Production and Significance of Lutein 7
        1.3.2.2 Bioactivities of Lutein 8
    1.3.3 Zeaxanthin 9
        1.3.3.1 Microalgal Production and Significance of Zeaxanthin 9
        1.3.3.2 Bioactivities of Zeaxanthin 9
    1.3.4 β-Carotene 11
        1.3.4.1 Microalgal Production and Significance of β-Carotene 11
        1.3.4.2 Bioactivity of β-Carotene 11
1.4 Biological Safety and Emerging Commercial Scope of Various Algal Pigments 12
1.5 Factors Affecting Carotenoid Production 13
    1.5.1 Effect of Important Factors during the Upstream Process in Bioreactors 13
    1.5.2 Effect of Other Factors during Growth in an Outdoor Environment 14
        1.5.2.1 Heavy Metals 14
        1.5.2.2 Pesticides 14
    1.5.3 Effect of Extraction Solvent on Carotenoid Recovery during Downstream Processing 15
1.6 Global Carotenoid Market 16
References 16

1.1 INTRODUCTION

The oceans encompass over 70% of this planet and are home to a diverse range of organisms, including marine algae. Algae are a diverse category of (mainly) photoautotrophic organisms that include both prokaryotic and eukaryotic species. They are divided into various taxonomic groupings, including Rhodophyceae (red algae), Chlorophyceae, Phaeophyceae, Xanthophyceae, Cyanophyceae, Bacillariophyceae (diatoms), and Dinophyceae (dinophytes or dinoflagellates).

Algae were used worldwide as a traditional food source in ancient times due to their nutritional value, diversity, distribution, and accessibility. However, algae are naturally rich in dietary fibers, proteins, minerals, polyunsaturated fatty acids, and vitamins. Moreover, they are also valuable sources of bioactive compounds such as polysaccharides, natural pigments, and phenolic compounds. Among these, algal pigments are in great demand due to their several health benefits without any side effects; hence Western cultures' interest has grown remarkably in their consumption in the recent past.

Algae (microalgae and macroalgae) are identified as natural pigment-accumulating organisms. They possess a huge pigment variety and are classified into three main groups: carotenoids, chlorophylls, and phycobilins or phycobiliproteins (Pereira et al., 2021), and they help to group algae into distinct phyla because they are the most frequent natural pigments in Ochrophyta, Chlorophyceae, and Rhodophyceae, respectively. Phycobiliproteins are water-soluble pigments also possessing proteins covalently bonded to tetrapyrroles called phycobilin's. They're used as light-gathering pigments in phototrophic systems. The protein content of phycobilins from blue-green algae, (*viz. Spirulina*), and Rhodophytes signifies up to 50% of the total cellular proteins. Based on spectral properties, phycobiliproteins are divided into four larger categories: Phycoerythrins (red, highly abundant), allophycocyanin (bluish-green), phycocyanin (blue), and phycoerythrocyanin (purple) (Catarina et al., 2020). Phycocyanin is the most frequent pigment in Cyanobacteria or blue-green algae.

All algae contain different types of basic photosynthetic pigments. Chlorophyll is a green fat-soluble pigment found in plants, algae, and cyanobacteria that takes part in photosynthesis. Chlorophylls contain porphyrin rings bound to a magnesium atom and facilitate light conversion into biological energy. They are the most important photosynthetic pigment. So far, five kinds of chlorophylls have been identified, chlorophylls a, b, c, d, and f, which were named according to the discovery order. Out of them, Chl a is the major and most important pigment participating in photosynthesis. In most cases, these are essential parts of the chloroplast lamella, but in cyanobacteria, they are evenly distributed throughout their protoplasm, known as chromatoplasm. Chlorophyll a is the most common pigment in all phototrophic cells. The only exception is phototrophic bacteria possessing bacteriochlorophyll. Chlorophyll a is common not only in higher plants but also in green, brown, and blue-green algae. Its yellow-green color can be hidden by the colors of other pigments. Chlorophyll a is the principal photosynthetic pigment in all phototrophic algae.

Moreover, other than chlorophyll, there are various pigments present in algae with various beneficial *in vivo* and *ex vivo* biological activities and remarkable bioavailability such as anticancer, antioxidant, anti-obesity, anti-inflammatory, anti-angiogenic, and neuroprotective activities (Patel et al., 2022). These carotenoids are mainly orange, red, and yellow natural pigments synthesized by microalgae and exert several vital physiological functions. For instance, in photosynthetic organisms, carotenoids are essential for *in vivo* application in photosynthesis and photooxidative protection, whereas after extraction, they exhibit these effects upon ingestion in the host.

Besides several health benefits, natural pigments have attracted a lot of attention in recent years for their aesthetic value (Azeem et al., 2019), Moreover, they can also be used as substitutes for artificial dyes/colorants (Azeem et al., 2019). Primarily, they have been exploited by the food and nutraceutical industries and later by the pharmacological industry. Microorganisms have been used in the food, medicinal, cosmetic, and energy industries for years. Among them, algae (microalgae and macroalgae) have proved a valuable source of beneficial compounds (Patel et al., 2021a, 2021e). Pigment yield is always proportional to the biomass yield; hence cost-effective extraction of pigment was achieved by advanced biomass cultivation techniques (Sim et al., 2019; Patel et al., 2019) and production design for higher pigment accumulation in biomass (Patel et al., 2022). Low production and processing costs render them even more attractive options for commercial-scale biological pigment production.

Mixotrophy is a higher biomass-yielding cultivation strategy that also offers a higher product yield (Patel et al., 2021a, 2021e). Recent efforts have further improved its performance, mainly through the selection of potential strains (Sung et al., 2019), improving light conditions (Hong et al., 2019b), bioreactor design (Sim et al., 2019), and inorganic carbon concentration (Hong et al.,

2019a; Patel et al., 2021d). In recent years this platform was highly explored to produce high-value algal products such as carotenoids (Patel et al., 2022), polysaccharides (Patel et al., 2013, 2021f), and oligosaccharides (Patel et al., 2021c). The traditional method of autotrophic microalgae cultivation was a low-yield platform that was mostly used for selective high-value product synthesis as well as tertiary treatment of wastewater (Perez-Garcia and Bashan, 2015). Advanced mixotrophic cultivation is a high-yielding platform that enhances the scope of biofuel and other low-value product biorefining. Moreover, it also opens up new methods for cost-effective algae as well as bioremediation scope (Patel et al., 2020a, 2020b, 2021b).

In this chapter, the growing scope of algal carotenoids for several health benefits for human consumption is summarized.

1.2 DE NOVO SYNTHESIS OF ALGAL PIGMENTS

The basic production of de novo microalgal pigment starts from the mevalonate (MVA) or Methylerythritol phosphate (MEP) pathway with a five-carbon-membered precursor molecule termed isopentenyl diphosphate (IPP) and dimethylallyl diphosphate (DMAPP). The top of Figure 1.1 illustrates the sequential steps of production of various pigments from the previous pathways via polyprenyl pyrophosphate chain elongation catalyzed by prenyl-transferase via continuous IPP condensation into DMAPP, extending the chain of polyprenyl pyrophosphate. The primary step towards the first precursor (phytoene) of this pathway is 20-carbon-membered geranylgeranyl diphosphate (GGPP) synthesis, catalyzed by GGPP synthase from two GPP (C10) condensation. During this carbon chain elongation, GGPP (C-20) converts into phytoene (C40) molecules. This pathway shows that the algal pigments are commonly C-40 carbon molecules with variability in their structure (mainly the number of H and O atoms) and activities. Popular carotenoids astaxanthin, zeaxanthin, and lutein are either intermediate or end products of the pathway. Besides the previously defined algal pigments, other known carotenoids are canthaxanthin, beta carotene, violaxanthin, voucherixanthin, loroxanthin, antheroxanthin, fucoxanthin, neoxanthin, diadinoxanthin, diatoxanthin, and so on (refer to Figure 1.1). All these carotenoids possess different degrees of bioactivity, such as

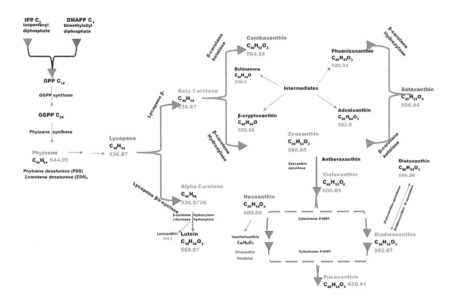

FIGURE 1.1 Mevalonate pathways and methylerythritol-4 phosphate pathway of microalgae illustrating various carotenoid production (Patel et al., 2022)

antioxidant, anti-obesity, anti-carcinogenic, anti-angiogenic, anti-inflammatory, and neuroprotective properties (Patel et al., 2022).

1.3 IMPORTANT CAROTENOID STRUCTURE, CLASSIFICATION, AND SIGNIFICANCE

Several pigments have the distinctive sub-atomic design of elongated carbon chains or aromatic rings bonded by conjugated double bonds. These bonds are mainly stable since they include resonance, a state in which more than one molecular structure can exist at the same time. For example, benzene oscillates between three and four classic structures. Adding oxygen or nitrogen to such a ring system raises the number of probable resonance sites. Simple heterocyclic rings of pyrrole are one of the basic components of many biochemical and organic pigments.

As discussed earlier, among three groups of pigments, carotenoids are the second most quantitatively available (Pereira et al., 2021). Carotenoids are tetraterpene pigments dispersed in some species of phototrophic bacteria and other species of fungi, archaea, plants, algae, and animals. Moreover, they are found in orange, yellow, red, and purple colors. As of 2018, approximately 850 naturally occurring biological carotenoids have been described. Carotenoids, along with chlorophyll, are vital pigments found in phototrophic tissues. Carotenoids also function as sunscreens, color-catchers or attractants, antioxidants, and precursors of some plant hormones in non-phototrophic tissues. As animals are unable to synthesize these carotenoids, animals bearing carotenoids are usually accumulated from food sources and/or moderately modify natural carotenoids via metabolism.

In general, carotenoids are 40-carbon skeleton color compounds. Out of them, a few are specific carotenoids bearing 45 or 50 carbon skeletons hence categorized as higher carotenoids, whereas carotenoids consisting of less than 40 carbon atoms in their carbon skeleton are termed apocarotenoids. These are found in a few plant and animal species as processed products of C40 skeleton carotenoids (Maoka, 2020). Carotenoids can be separated into two groups: carotene and xanthophylls. In nature, there are about 50 types of carotenes, with main hydrocarbons such as α-carotene, β-carotene, γ-carotene (ψ-carotene), and lycopene. Carotenoids play crucial roles in animals, for example, acting as vitamin A precursors, photoprotectors, and antioxidants, boosting immunity and contributing to reproduction, whereas xanthophylls, for example, lutein, β-cryptoxanthin, zeaxanthin, fucoxanthin, astaxanthin, and peridinin, are natural carotenoids comprising one or more oxygen atoms as carbonyl, hydroxy, carboxyl, aldehyde, furanoxide, and epoxide groups in these compounds. As of 2018, about 800 species of xanthophylls have been reported in nature. Some of the important xanthophylls are described in more detail in the following.

1.3.1 Astaxanthin

1.3.1.1 Microalgal Production and Significance of Astaxanthin

Astaxanthin is a xanthophyll responsible for the reddish color of salmon, trout, and crustaceans. It is also used as poultry feed to get coloration of the tail and flesh. Synthetic astaxanthin is a mixture of three isomers. Astaxanthin from *Haematococcus* is exclusively a (3S, 3'S) isomer and is mostly esterified with fatty acids. Astaxanthin (red-orange carotenoid) is the most commercially significant carotenoid. It is widely used in the nutraceutical, pharmaceutical, cosmetic, feed, and food industries. The astaxanthin market is the third largest (after lutein and β-carotene) world carotenoid market, reaching $288.7 million in 2017, and is projected to grow to $426.9 million by 2022. A significant rise in astaxanthin demand is anticipated in the coming years. The price of natural astaxanthin is much higher, at 15,000 US $/kg, than any synthetic astaxanthin (1000 US$/kg) (Hu, 2019). Astaxanthin can be used as a nutritional supplement that inhibits oxidation and free radical formation in the metabolism; stimulates immunization; and prevents ulcers, gastric injury, cardiovascular disease, diabetes, and neurodegenerative disorders as well as cancer (Sharma et al., 2019).

Marine Algae as a Source of Carotenoids

Among many species of microalgae, one of the most important and popular strains for xanthophyll production is *Haematococcus pluvialis*, which is one of the strongest natural antioxidant sources for astaxanthin. It was found to be 65 times stronger in antioxidant activity than vitamin C and 10 times stronger than other carotenoids (Patel et al., 2022). It accumulates up to 3.8% astaxanthin on a dry-weight basis (Sharma et al., 2019). A maximum of >13 mg/L/day productivity and >40 or >190 mg/L yields have been achieved from *H. pluvialis* within 13 days of cultivation. The *Haematococcus lacustris* strains 26 and WZ were reported for astaxanthin yields of 51 and 40 mg L^{-1}, respectively, equivalent to 2.8 and 2.5% of their dry biomass. This microalga has gained popularity as a commercial model organism for astaxanthin which has wider health applications. In the last decades, a large number of publications have described the potential use of *H. pluvialis* in various algae biorefining schemes as well as its high-value molecule astaxanthin.

H. pluvialis–derived astaxanthin is better in quality and efficiency; however, the yield is lower, and cultivation time is prolonged. *H. pluvialis* is not a potential strain for mixotrophic cultivation, as it is unable to utilize all forms of organic sugars such as glucose, xylose, fructose, and sucrose. Some studies reported its weaker mixotrophic growth in acetate and alcohol-containing media. Researchers are looking for alternative strains with higher mixotrophic potential than *H. pluvialis* that can also support biofuel production from residual biomass and lipid fraction (Choi et al., 2019). *Chromochloris zofingiensis* was found to be interesting under this scheme; however, astaxanthin bioactivity is lower than that of *H. pluvialis*, but its yield can be significantly enhanced using a strategic implementation of mixotrophic cultivation (Patel et al., 2019) or a combination of switching autotrophic and mixotrophic modes as applicable for yield increase (Sim et al., 2019). So far with this strain, a maximum <3 mg/g yield within 8 days of cultivation (Sun et al., 2019) and 6 mg/g within 10 days (Kou et al., 2020) have been reported. Astaxanthin production from flue gas found to be a cost-effective strategy that has been approved for safe human consumption recently (Hong et al., 2019a). With its bioactivity such as anti-inflammatory, anti-aging, anti-hypertensive, and anticancer, astaxanthin is largely used in pharmaceuticals, functional foods, and cosmetics as a free radical scavenger (Patel et al., 2022).

Biotech company Micro Gaia, an industry leader in the production of astaxanthin-rich microalgae under nitrogen limitation, proposed using moderate nitrogen limitation as one step in the continuous manufacturing process. Microalgae *H. pluvialis* is not efficiently cultivated in dark heterotrophic and efficient mixotrophic modes, so astaxanthin production largely manifests in phototrophic mode, which is dependent on light irradiance, which is a cost factor. A green microalga called *C. zofingiensis*, which is robust in dark and mixotrophic growth, easy to cultivate, and tolerant to environmental fluctuations has been introduced as a possible alternative for astaxanthin production: it grows three times faster than *H. pluvialis* and stores substantial amounts of biomass and carotenoids in the dark, therefore facilitating large-scale economic production of astaxanthin. *C. zofingiensis* contains up to 6.5 mg of astaxanthin per gram of dry weight. It has been attracting a great deal of attention due to its diverse capability to produce lipids, carotenoids, and exopolysaccharides under fluctuating growth conditions. Table 1.1 summarizes the astaxanthin production profile from microalgae with details of the contents and productivity.

1.3.1.2 Bioactivity of Astaxanthin

Astaxanthin is identified for several pharmacological activities such as immunoprophylactic, antidiabetic activity, antioxidative, anti-inflammatory, and anticancer. The antioxidant efficacy of astaxanthin is several times that of other natural antioxidants, such as α-tocopherol, Trolox, lycopene, zeaxanthin, lutein, and α/β-carotene, (Yeh et al., 2016). Astaxanthin's role in the prevention of diabetes has often been reported. Commonly, patients with diabetes suffer from hyperglycemic oxidative tissue injury recovered by astaxanthin application. A 2-mg kg^{-1} d^{-1} astaxanthin dosing rate for four weeks exhibited a significant anti-hyperglycemic effect (Bhuvaneswari et al., 2014). Astaxanthin suppresses body weight, blood glucose, and homeostasis, concurrently increasing

TABLE 1.1
Astaxanthin and Lutein Production and Extraction Yield from Various Microalgae Cultures

Pigment	Algae Source	Cultivation Mode	Production Yield	Extraction Method	Extraction Yield
Astaxanthin	*Haematococcus pluvialis*	—	4 g biomass extract	Cell disruption, monoester fractionation, and solvent used	148 mg biomass
	Arthrospira platensis	Phototrophic	2.03 g L^{-1}	Sonification and solvent used	0.25–0.60 mg/g DW
	Dunaliella viridis	Phototrophic	45.00 ± 4.27 × 10^5 cells/mL on day 14	Cell disruption (liquid nitrogen) and solvent used	77.5 ± 7.7 µg/g
	Coelastrum sp.	Phototrophic	1 × 10^6 cells/mL	Solvent extraction	28.32 ± 2.5 mg/L
	Chromochloris zofingiensis	Phototrophic	0.6 g/L in culture media on day 6	Solvent extraction	2.8 mg/g
Lutein	*Scenedesmus* sp.	Mixotrophic	—	Microwave-assisted, KOH 60% Solvent: acetone	11.92 mg/g
	Chlorella vulgaris G-120	Heterotrophic	—	Ultrasound-assisted n-heptane–ethanol–water	3.20 mg/g
	Chlorella sorokiniana (NIOT-2)	Autotrophic	—	Microwave-assisted extraction cell disruption using KOH and l-ascorbic acid	20.69 ± 1.2 mg/g
	Parachlorella kessleri HY1	Heterotrophic	22kg/1000L	n-heptane–ethanol–water, 5:4:1.5 Maceration extraction (ME) 30 min	6.71 mg/g

serum astaxanthin levels and quantitative sensitivity of insulin. Moreover, treatment with it significantly increased hepatic glycogenesis in diabetic hosts while inhibiting gluconeogenesis (Bhuvaneswari et al., 2014). The oral application of 2 mg kg^{-1} d^{-1} astaxanthin was effective against lipid accumulation, reactive oxygen species (ROS) production, caspase 12 activation, and stress markers of the endoplasmic reticulum. Moreover, it was also reported for activating 2α phosphorylation factor, the sixth transcription factor in high fructose-fat diet-fed diabetic mice (Bhuvaneswari et al., 2014). Astaxanthin effectively inhibits diabetic inflammation in mice (Bhuvaneswari et al., 2014).

Peroxisome receptors (PPAR-γ) play a role in carbohydrate metabolism and it is a prime receptor of astaxanthin. Astaxanthin binds with PPAR-γ and facilitates its communications with TIF2 and SRC-1 receptors prior to regulating PPAR-γ gene synthesis. The 50 mg kg^{-1} d^{-1} of astaxanthin treatment was able to reduce elevated glucose in diabetic rats (Park et al., 2015). Moreover, astaxanthin attenuates the levels of ROS, glycation, and lipid peroxidation; however, no change was found in cellular antioxidation or in ROS-induced proteins (Park et al., 2015).

A month's application of 20 mg kg^{-1} d^{-1} of astaxanthin exhibited the lymphocyte's redox status in diabetic rodents. Moreover, astaxanthin responds to redox imbalance by controlling ROS/RNS metabolism. Astaxanthin also enhances catalase, GPx, GR, and SOD activity in the lymphocytes of diabetic rats. Hyperglycemia-induced apoptosis or even inflammation in the central nervous system leads to diabetic neuropathy (Dewanjee et al., 2018). A 10–40 mg kg^{-1} d^{-1} astaxanthin application delayed cognitive deficiencies and neurotoxicity (Xu et al., 2015). Moreover, the enhanced level of NF-κB (p65), NOS, TNFα, ILs, and apoptotic attributes were found in the hippocampus and cerebral cortex of diabetic rats with type 1 (Xu et al., 2015). These results confirmed that astaxanthin possesses bioactivity that makes it a potential pharmaceutic and functional food (Dewanjee et al., 2018).

By improving cell glutathione levels, inhibiting the production of malondialdehyde, and restoring endogenous SDS levels, astaxanthin showed neuroprotective properties when used in H$_2$O$_2$-induced neurodegeneration. Neuronal death and injury are inhibited by astaxanthin through the activation of ERK1/2 pathways. It modulates glutathione-S-transferase-1, heme oxygenase, and quinone oxidoreductase levels to inhibit oxidative stress in the body. Astaxanthin decreases brain aging by enhancing SOD, GPX, and GSH activity and reducing DNA adducts, MDA, and carbonylation protein. A previous study showed that by increasing nitric oxide bioavailability against ROS, astaxanthin had antihypertensive and antithrombotic effects. Astaxanthin derived from *Haematococcus lacustris* mitigates the effects of UV exposure on melanocytes, assisting in preventing skin cancer.

1.3.2 LUTEIN

1.3.2.1 Microalgal Production and Significance of Lutein

A natural antioxidant, lutein has attracted tremendous consumer attention for its health-promoting properties. It has been found that lutein intake at a recommended dose (6 mg day^{-1}) is beneficial for human health because the human body cannot synthesize it. Among its potential uses for skin rejuvenation, lutein also prevents Alzheimer's disease and age-linked macular degeneration (Ren et al., 2021b). The human retina accumulates lutein absorbed into the body, which filters blue light, protecting eyesight. Marigold is the major source for the extraction of commercial natural lutein. A wide variety of marigold species contain high levels of lutein, ranging from 17 to 570 mg per 100 g (Lin et al., 2014). However, the need to harvest them in specific seasons and time-consuming petal separation are major limitations. Lutein can also be found in leafy green vegetables, yellow fruits, broccoli, and egg yolks, but its concentrations and bioavailability are low.

Replacing marigolds with microalgae brings considerable economic benefits, bioactivity, bioavailability, and ease of the process, especially harvesting and extraction (Lin et al., 2014). Several potential microalgae have been examined for their lutein production potentials, such as *Chlorella, Scenedesmus, Parachlorella, Tetraselmis*, and *Coccomyxa* (Patel et al., 2022; Rajput et al., 2021).

For lutein production, temperature, pH, trophic modes, nutrient availability, and salinity affect it remarkably. Among various trophic modes, mixotrophy was most efficient to obtain a higher yield than other trophic modes (Patel et al., 2022). Moreover, other factors such as carbon source, light condition, and light and dark cycle also affect lutein yield and productivity. High irradiance is often inhibitory for pigment production involved in light capturing, whereas low light is more favorable, as microalgae are inclined to capture more light by synthesizing more pigments in low-light conditions. Microalgae such as *Chlorella minutissima* and *Chlorella sorokiniana* are able to efficiently assimilate glucose, acetate, and other organic forms (Patel et al., 2022). Temperature fluctuation and its effects were closely associated with irradiance level; hence they must be studied combined.

The study showed that the lutein content of biomass is not significantly affected by nitrogen (nitrate) concentration in the culture medium; however, nitrogen limitation reduces biomass productivity, which ultimately reduces lutein synthesis. As such, nitrate should be supplied in a moderate excess to avoid growth inhibition but to avoid saline stress, which can adversely affect culture performance (Sánchez et al., 2008).

The lutein yields and productivity from *Chlorella* sp. with 20 g L^{-1} glucose respectively offered 10.5 mg L^{-1}d^{-1}, 63 mg L^{-1} which was >7.3 mg L^{-1}d^{-1}, and 44 mg L^{-1} with 10 g L^{-1} glucose (Wang et al., 2020b). An efficient production rate of 8.25 mg L^{-1}d^{-1} has also been achieved with two-phage cultivation, for example, the mixotrophic-phototrophic process of *C. sorokiniana* (Wang et al., 2020b). With sodium acetate, *Chlorella sorokiniana* Kh12 can accumulate >13 mg g^{-1} DW of lutein (Patel et al., 2022). Various wastewater and waste products have been utilized for microalgae cultivation for the production of lutein and lipid under biorefinery approaches, such as poultry litter, aquaculture wastewater, and food waste (Ren et al., 2021a). Table 1.1 summarizes the results of lutein production from microalgae with details of the contents and productivity.

1.3.2.2 Bioactivities of Lutein

Glucose-induced oxidative stress and inflammation can be constrained by lutein both *in vitro* and *in vivo* in immune system cells (Muriach et al., 2008). Similar to lutein, the research disclosed zeaxanthin's defensive role in diabetic retinopathy. Lutein helps to control hyperglycemia-induced high lipid peroxidation, GSH level reduction, and NF-κB signaling activation in high glucose-induced cells (Muriach et al., 2008). This shows that lutein can reduce oxidative stress and impairment of immunity irrespective of hyperglycemia in diabetics (Muriach et al., 2008). Besides inhibiting NF-κB, lutein also downregulates inflammatory species, MCP-1, ICAM-1, and FKN (Yeh et al., 2016). Lutein delayed diabetic retinopathy development. A 6 mg d^{-1} lutein intake for three months improved contrast sensitivity, visual acuity, and macular edema under diabetic retinopathy. Lutein consumption also reduces the onset of cataracts in diabetics with type 2. Lutein with and without insulin combination reduces lipid peroxidation and enhances GSH concentrations in diabetic rats with type 1 (Arnal et al., 2009). Thus, lutein can treat diabetic retinopathy via redox defense and prevent retinal inflammation by blocking MCP-1, ICAM-1, FKN, and NF-κB (Yeh et al., 2016).

Very low doses of lutein can hamper lipid peroxidation and increase GPx and GSH levels in diabetic mice at unaltered glycemia (Muriach et al., 2006). Moreover, it can stop retinal NF-κB activation in diabetic mice with type 1 (Muriach et al., 2006). Thus, lutein is also being used as an adjunct therapy to diabetic retinopathy via redox protection and suppressing retinal inflammation (Muriach et al., 2006). Lutein saves diabetic kidneys through mechanisms of oxidative protection. Lutein has been found to play a fundamental role in synaptic binding by increasing cell availability in the nervous system, with direct/indirect roles in the gene expression of binding proteins. Oxidative pressure and inflammation have inevitably been thought to take part in pathological brain change during DM (Muriach et al., 2006).

Twelve-week treatment with 0.5 mg d^{-1} lutein can significantly stop lipid peroxidation and increased levels of GSH and GPx at the cerebral cortex of diabetic rats without upsetting glycemia (Arnal et al., 2010). However, the statistical analysis of the report seems unconvincing (Arnal et al., 2010). 4-HNE immunofluorescence analysis in diabetic rats (type 1) maintained the defensive lutein role for diabetic encephalopathy (Arnal et al., 2010). In summary, lutein may serve as an adjunct therapy for diabetic microvascular problems. *Chlorella*-derived lutein exhibited an anti-proliferative effect on a human cell line (HCT116) better than pure algal lutein. Moreover, *Chlorella*-derived lutein also inhibits apoptosis associated with an anti-proliferative effect (Martinez-Andrade et al., 2018). Lutein has been recognized as a potential macular carotenoid helping to scavenge reactive oxygen species generated from photooxidation reactions and thus reducing lipid, protein, and DNA damage. Moreover, it is protecting against apoptosis, and reduces light exposure to prevent photoreceptor activation (Mrowicka et al., 2022).

1.3.3 ZEAXANTHIN

1.3.3.1 Microalgal Production and Significance of Zeaxanthin

Human eyes and skin contain another xanthophyll pigment: zeaxanthin. Zeaxanthin accumulates, along with lutein, in the macular region of the cornea, protecting the retina from blue light and enhancing visual acuity. Furthermore, zeaxanthin also exhibits antioxidant and anti-inflammatory properties as well as being used to prevent neurological disorders. Zeaxanthin treatment improves cognitive impairment and memory function in consumers. Some vegetables, fruits, and cereals also contain a small concentration of zeaxanthin. The contents of cooked scallions, orange peppers, and cornmeal are relatively low in zeaxanthin, at 24.9, 16.7, and 16.3 µg/g, respectively (Ren et al., 2021a).

Zeaxanthin is currently produced commercially from marigold flowers. Zeaxanthin content varies from 10 to 300 mg g^{-1} in different marigold species. There is an attractive prospect of zeaxanthin production from microalgae, and the production rate is dependent on many factors, including the microalgae species cultivated (Sarnaik et al., 2018). Similar to lutein, zeaxanthin production is also affected by light irradiance intensity and period. Depending on the microalgae species, the optimum light condition for zeaxanthin production varies. By high light exposure, *C. zofingiensis* accumulated 5.69 mg g^{-1} DW of zeaxanthin in comparison to its low-light counterpart (0.34 mg g^{-1} DW) (Huang et al., 2018). The same strain was reported to accumulate higher zeaxanthin up to 7 mg g^{-1} DCW with high light and nitrogen limitation effects together. However, higher light is not always found to be favorable for zeaxanthin production. Zeaxanthin exhibits chemical structures similar to lutein. A previous study developed an efficient separation method of zeaxanthin and lutein by a YMC Carotenoid C30 column equipped with ultra-HPLC (Huang et al., 2018).

Zeaxanthin is increasingly in demand due to the rise in AMD sufferers and the increase in natural sources with higher productivity. The marine green alga *Chlorella ellipsoidea* produces nine times more zeaxanthin than a red pepper. In addition, zeaxanthin derived from microalgae is for the most part found as free zeaxanthin, dissimilar to mono- or di-esters present in plants. Therefore, microalgae are rather more beneficial for the production of zeaxanthin due to their higher productivity and bioavailability. Besides zeaxanthin production, its extraction needs attention to make its bioprocess Cost-effective. It includes a cheaper cell disruption method as well as a solvent to recover maximum zeaxanthin from microalgal cells (Mitra and Mishra, 2019). Advanced techniques such as pressure disruption, ultrasound, and microwave pretreatment before liquid extraction were highly effective to recover zeaxanthin from microalgal *Chlorella ellipsoidea* cells (Mitra and Mishra, 2019; Patel et al., 2022). The production and extraction profile of zeaxanthin from microalgae is summarized in Table 1.2 with details of productivity.

1.3.3.2 Bioactivities of Zeaxanthin

Zeaxanthin intake improves the consumer's cognitive impairment and memory function. Taking zeaxanthin at 50 mg kg^{-1} for 12 weeks improves cognitive impairment through glycemic control, redox defense protection, neuronal protection, and NFκB translocation inhibition in the hippocampus of high-sugar and fat–dosed rats with type 2 diabetes. A 200- and 400-mg kg^{-1} intake rate of zeaxanthin for four weeks in these rats showed normalized body weight and hyperglycemia (Kou et al., 2017). It also showed significant antihyperlipidemic activity in the rats by regulating low-density lipoprotein levels of cholesterol in serum, triglycerides, HDL cholesterol, and total cholesterol (Kou et al., 2017). Moreover, zeaxanthin played a prophylactic activity in diabetic nephropathy via returning renal physiology, glucosaminidase and albumin levels in urine, and urea nitrogen level in blood serum to close to normal (Kou et al., 2017). Zeaxanthin can effectively reduce hyperglycemic inflammation by dropping the levels of inflammatory factors IL6, IL2, NFκB, and TNFα (Kou et al., 2017).

Zeaxanthin exhibits remarkable antioxidant effects in rats with type 2 diabetes by increasing serum levels with antioxidant molecules catalase (CAT), superoxide dismutase, glutathione peroxidase,

TABLE 1.2
Zeaxanthin and Lutein Production and Extraction Yield from Various Microalgae Cultures

Pigments	Algae Source	Cultivation Mode	Production Yield	Extraction Method	Extraction Yield
Zeaxanthin	*Synechococcus* sp.	Autotrophic	4.25 ± 1.42	Freeze dried; high-impact zirconium beads added with methanol	3.32 ± 1.25 mg/g
	Porphyridium purpureum	Autotrophic	3.4 g/L	Frozen algal cells were ground in liquid nitrogen before resuspension in 0.1 M phosphate buffer (pH 6.8) for overnight	269 μg/g
	Nannochloropsis oculata	Mixotrophic	—	100% acetone; sonicated at 0 °C (ice-water bath);	48 ± 0.88 mg/L
	Synechococcus elongatus	Autotrophic	0.91 ± 0.06 g/L	Absolute methanol; centrifuged, resuspended in absolute methanol, and sonicated	9.02 ± 1.10 mg/g
β-carotene	*Scenedesmus quadricauda* PUMCC 4.1.40	Autotrophic	1.5 g/L	Extracted with acetone, manual shaking at 50°C	19.0 mg/g
	Chlamydomonas reinhardtii A1 strain	Mixotrophic	—	Vortexing with zirconium beads; extracted with ethanol	23.75 mg/g
	Coelastrella sp.	Autotrophic	—	Frozen under –80°C for 24; ground with liquid nitrogen in a mortar and pestle; methanol and dichloromethane (75:25 in v/v) utilized for the extraction	13.15 mg/g
	Tetraselmis suecica	Autotrophic	—	Sonication; extracted with acetone and methanol	1.0 mg/g

and methane dicarboxylic aldehyde. Obesity is one of the major difficulties in diabetic patients with type 2 and may also decrease insulin sensitivity (Sharma et al., 2017). Zeaxanthin inhibits lipogenesis in 3T3L1 adipocytes via regulating energy metabolism *in vitro* by 5'-AMP-induced protein kinase (AMPK) (Liu et al., 2017). A four-week intragastric treatment of 20 mg kg^{-1} zeaxanthin reduces lipid accumulation while inducing AMPK activation, reduces adipocyte size, and reduces fat weight in obese mice kept on a high-fat diet to test the anti-obesity effect of zeaxanthin (Liu et al., 2017). Zeaxanthin also suppresses fatty acid synthase, sterol regulatory molecule binding protein, and PPARγ in adipocytes (Liu et al., 2017).

Several recent studies clearly show that high lutein and zeaxanthin-based dietary intake significantly reduces early age-related macular degeneration, associated with genetic risk variants (Mrowicka et al., 2022). As both are strong antioxidants similar to astaxanthin with long-conjugated double bond molecules, they are efficient in scavenging reactive oxygen species induced during stress conditions due to high light intensity. Both lutein and zeaxanthin protect the retina and eye lens from age-related changes.

1.3.4 β-Carotene

1.3.4.1 Microalgal Production and Significance of β-Carotene

β-carotene is a red-orange carotenoid and is commonly found in pumpkins, mango, and carrots, as well as in fungi (*Phaffia rhodozyma*) and microalgae (Wang et al., 2021). It is a lipid-soluble carotenoid and acts as a precursor of Vitamin A in humans for night blindness treatment. It was the first natural high-value product from microalgae ever commercialized. Its antioxidant, anti-cardiovascular, and immune-boosting properties make β-carotene widely used in food and medicine (Wang et al., 2021). Despite being inexpensive, chemically synthesized β-carotene is less appealing to consumers than natural β-carotene. Besides β-carotene's application as a food colorant, it is an active ingredient in the pharmaceutical and cosmeceutical industries. The price range of natural β-carotene is approx. 300–3000 USD/kg, based on purity (Ebrahimi Mohammadi and Arashrad, 2016).

Dunaliella salina is identified as the main source of natural β-carotene, and it can accumulate β-carotene up to 10–13% DW when induced by high temperature, high light, nutrient deprivation, and high salinity (Lamers et al., 2012). As *D. salina* can grow in high salinity, its cultivation in seawater offers a high yield of β-carotene and thus more profits. Moreover, other species of microalgae also reported synthesizing β-carotene. *Arthrospira platensis*, *Chlorella zofingiensis*, and *Caulerpa taxifolia* accumulate β-carotene up to 0.1–2% DW. There have been studies that show spirulina contains ten times the amount of β-carotene as carrots (Wang et al., 2021). Under high light and salt stress, a mutant *C. zofingiensis* bkt1 could accumulate high content (mg g^{-1} DW) of three carotenoids: lutein-14, β-carotene-7, and zeaxanthin-7. The production profile of β-carotene from microalgae is summarized in Table 1.2 with details of productivity. Other β-carotene–producing microalgae are *Eustigmatos polyphem*, *E. magnus*, *E. vischeri*, *Vischeria punctata*, *V. helvetica*, and *V. stellate*. Some beta carotene–producing bacteria are also reported, such as *Sphingomonas sp.* and *Serratia marcescens*, able to produce beta carotene at 29 % (of biomass dry weight) and 2.5 mg ml^{-1}, respectively.

1.3.4.2 Bioactivity of β-Carotene

β-carotene derived from microalgae is a potent antioxidant agent. β-carotene intake has been shown to reduce the likelihood of emerging diabetes in both women and men. There is an opposite association between plasma β-carotene and plasma glucose during insulin resistance as well as fasting. Therefore, low blood β-carotene levels are detected in line with decreased insulin sensitivity. Many studies explained an opposite correlation between intake of beta carotene and the chance of type 2 diabetes. Diabetic retinopathy is the main microvascular problem and causes symptomatic impairment. Retinopathy is mainly caused by insistent hyperglycemia. β-carotene plays an important role in both vision and macular pigment recovery. β-carotene acts as provitamin A and is found in the ocular tissue. People with diabetes are more likely to develop cataracts, leading to poor vision and full blindness (Krinsky and Johnson, 2005). β-carotene has been shown to inhibit the onset and progression of cataracts through antioxidant mechanisms. A population-based study shows a significant inverse correlation between blood β-carotene levels and cataract development from 11 villages in northern India (n = 1112, age 50 and older). However, β-carotene showed an opposite correlation with body mass index (P = 0.01). Moreover, β-carotene may offer added benefits by converting into vitamin A, an important functional component of the human retina (Krinsky and Johnson, 2005). Key carotenoids from algal sources and their bioactivities are summarized with their possible applications in Figure 1.2.

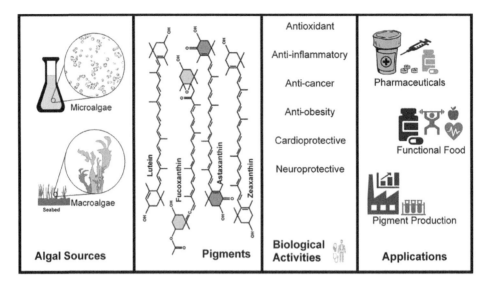

FIGURE 1.2 Important algal carotenoids, their bioactivities and health applications

1.4 BIOLOGICAL SAFETY AND EMERGING COMMERCIAL SCOPE OF VARIOUS ALGAL PIGMENTS

Numerous macro- and micro-algae are enriched in carotenoids, and they basically help in the absorption of sunlight at the growth stage. It is likely that carotenoids defend cells from oxidative radicles by quenching singlet oxygen injury through a variety of protective mechanisms. As such, carotenoids that have been derived from microalgae could be a strong green resource in the search for good functional ingredients. These carotenoids are also used as food pigments industrially in dairy products and beverages, among others, as feed additives. They also find application in cosmetics and in pharmaceuticals given the growing demand for natural products in these sectors. Some carotenoids also provide provitamin A. There has been an increase in interest in dietary carotenoids due to their ability to reduce the incidence of some chronic diseases that involve free radicals and their antioxidant properties after appropriate dosing.

Many steps are being considered for screening, selection, production, and applications of microalgae pigments. Furthermore, synthetic pigments from microalgae have dominated current commercial markets, particularly astaxanthin, with a total value of $200 million. It shows the growing market demand for microalgae-derived natural pigments for commercial applications as nutraceuticals, and the cultivation of microalgae on an industrial scale has extreme potential and an impressive market. The scale of the nutraceutical pigments market is rising; cultivation and extraction from *Haematococcus pluvialis* have been successfully produced at a commercial scale with natural pigments worldwide.

Microalgae pigments have diverse biological functions and may have a great impact on human health. The biological safety of pigments from diverse sources of microalgae has been evaluated by some authorities, including the European Food Safety Authority (EFSA), World Health Organization (WHO), and Food and Drug Administration (FDA), for their possible use as food supplementation or food ingredient. Purified astaxanthin derived from *Haematococcus pluvialis* has been approved for human intake, especially as a dietary supplement (at dosages up to 12–24 mg per diet for no more than a month) as well as by EFSA and FDA. At the same time, pigments from *Chlorella* species are often sold as "healthy foods" and are being used to improve the condition of communal or acute diseases like Alzheimer's and cancer. The WHO has approved the genus *Arthrospira* due to growing concerns about healthy or nutritious foods. It can increase the hemoglobin levels of red blood cells in elderly persons and strengthen the immune system. Moreover, it is recommended to add this to NASA astronauts' diet in space.

1.5 FACTORS AFFECTING CAROTENOID PRODUCTION

1.5.1 Effect of Important Factors during the Upstream Process in Bioreactors

For better commercial prospects of algal carotenoids, the first step of improvement lies in enhancement of algal biomass yield. Algal biomass yield is directly proportional to carotenoid yield. For economic production of microalgal biomass, fast-growing and high carotenoid–yielding microalgae strains must be adopted before developing a bioprocess. The key parameters which greatly affect microalgal biomass production during the growth stage are light intensity, light and dark cycle, temperature, pH, salinity, nutrient level, nutrient mixing rate and gas purging rate. Among these factors, light condition is very important for microalgae growth, which greatly varies depending on the microalgae strain and its range of light requirements (especially light intensity and duration) for photosynthesis. The study showed varying wavelengths of light have a great effect on the pigment concentration of *Chlorella pyrenoidosa* in photosynthesis. The concentration of chlorophyll-a decreases under blue light; however, it was considerably improved under mixed blue and red light. Moreover, the concentration of chlorophyll b was significantly reduced under red light, whereas no significant variation in its content was found with other light wavelengths (Wang et al., 2020a). The content of chlorophyll a and carotenoid gradually decreased in *Pyropia haitanensis* by 23.95% and 41.21%, under blue light and 22.76% and 37.90% under green light (Wang et al., 2020a). Homogenous distribution and availability of nutrients to microalgal cells is a very important parameter, especially during the growth phase for biomass and carotenoid accumulation. Thus, the role of mixing is crucial to avoid nutrient limitations in their microsphere environment once consumed. Likewise, adequate CO_2 purging is important to maintain a desirable inorganic carbon concentration in the liquid stream to attain the optimal growth rate; however, an excess flow rate affects the economy and limits the CO_2 flow rate, limiting the photosynthesis rate and growth.

Temperature is also an important factor that affects the production of biomass in microalgae. It mainly affects the metabolic performances, structures, and properties of microalgae cell envelopes. High temperatures were found to imbalance hydrophobicity and enhance carotenoid synthesis in certain microalgae (Novosel et al., 2022). High temperatures can lead to the production of many carotenoids in blue-green microalgae culture, including β-carotene. The ideal temperature for microalgae growth will depend on the strain and the microalgae's resistance to temperature changes. It was found that 22°C is the optimum temperature for *Dunaliella salina* growth, leading to the highest levels of β-carotene production than other temperature ranges. A desirable pH range between 5.5 and 8.5 is tolerable for the majority of microalgal survival and growth; however, close to a neutral or slightly alkaline pH range is found to be the best condition for optimal growth of microalgae. For example, *Nannochloropsis* sp. and *Tetraselmis* sp. achieve higher biomass and higher carbohydrate, protein, and lipid content between pH 7.5 and 8.5. *Arthrospira platensis* exhibited the maximum content (in mg/g DW) of carotenoid (2.4), chlorophyll a (10.6), phycobiliprotein (91), and phycocyanin (91) at pH 8.5, whereas it produced greater phycobiliprotein (159 mg/g DW) at pH 9.0, which clearly shows the pH effect on various carotenoid accumulation. These two parameters, pH and temperature, have a greater role during the extraction of microalgal carotenoids to enhance solubility and reactivity for their enhanced recovery from the cellular structure.

Salinity has a notable effect on algal pigment production. Salinity has a direct role in cellular osmosis regulation in marine microalgae. Salinity forms a hypertonic solution intracellularly at high concentrations of sodium chloride in the external environment. Therefore, cells contract when cell dehydration-induced damage occurs in internal components (Lopez and Hall, 2021). The fucoxanthin yield of the marine microalgae *Chaetoceros muelleri* at salt <55‰ and *Amphora* sp. at >55 ‰ salinity was respectively 2.92 and 1.2 mg g^{-1}, which shows the varying level of salinity requirements depending on the microalgae strain. The combination of factors was also studied in previous work. Under salt stress and light deficiency, *Chlorella zofingiensis* accumulates better canthaxanthin as compared to astaxanthin. This shows that light is not a limiting factor for canthaxanthin production, but adequate light availability is vital for astaxanthin production.

1.5.2 Effect of Other Factors during Growth in an Outdoor Environment

Organic and inorganic pollutants such as heavy metals and pesticides have great effects on algae growth and carotenoid production when microalgae grow in natural outdoor environments.

1.5.2.1 Heavy Metals

Human industrial development has created conditions for toxic wastes to leak into the environment. Only small concentrations, including cobalt, copper, and zinc, can affect the outcome of organisms. Several microalgal species, especially the green ones, are presently utilized for biological purification of industrially polluted environments containing heavy metal–contaminated areas. Green algae can modulate polluted environments via metabolic activity. Therefore, the studied effects of adding copper (Cu^{2+}), cadmium (Cd^{2+}), and lead (Pb^{2+}) in the culture medium of *Scenedusmus incrassatulus* revealed that Cd and Pb are highly toxic, severely inhibited growth, reduced pigment content, and increased lipid peroxidation. At the same time, Cu did not show significantly inhibited growth and chlorophyll biosynthesis in algal species (Purbonegoro et al., 2018).

A study was carried out on the unicellular blue-green algae *Synechocystis* sp. to find the effect of heavy metals (Pb^{2+} and Cd^{2+}) on their growth, structure, and pigment quantity. The results showed that increasing the concentration of heavy metals will generally reduce growth, cell density, biomass, and growth rate. The significant growth inhibition of *Synechocystis* sp. was determined in Cd^{2+} supplemented media than the Pb^{2+} supplemented media indicating that the former is more toxic. To compare the effects on algae growth in the presence or absence of heavy metals Cu, Co, and Zn were studied in the green algae *Chlorella vulgaris*. At metal concentrations of more than 10^{-9} M, growth rates decrease.

Besides, adding heavy metal content at an appropriate level during algae growth will increase growth performance. A lower CO_2 concentration (0.01 ppm) stimulated the growth of *Nostoc muscorum*. But when the concentration of heavy metals increases, it gives the opposite result. When adding 10 μM concentrations of Cu, Pb, and Cd, the growth of *C. vulgaris* decreased by 21%, 36%, and 53%, respectively (Piotrowska-Niczyporuk et al., 2012).

1.5.2.2 Pesticides

One prevalent kind of pollution that has a profound impact on marine environments is pesticides. Annually, consumption is about 3 million tons, and pesticides are among the world's most-used chemical compounds. Pesticide environmental contamination is a concern, as most are highly toxic to many targets and non-target organisms, including microalgae. A study by Perales-Vela et al. (2021) indicated the impact of methyl viologen on the total chlorophyll and carotenoid content of *Chlorella vulgaris*. Total chlorophyll and carotenoid concentration declined by 57.5% and 46.4%, respectively, as it was exposed to 2.5 M of MV for 72 hours. According to a recent study, the chlorophyll-a and chlorophyll-b content of freshwater alga *Chlorella* sp. was significantly reduced by 53.50% and 48.00%, respectively, due to its exposure to alpha-cypermethrin (ACy). Flazasulfuron ns led to a decrease in the chlorophyll pigment of *Scenedesmus obliquus* at concentrations of 10 µg/L and even at the lowest tested concentration, about 0.1 µg/L.

Other important factors affecting carotenoid yield are extraction solvents and their efficiency during the downstream process.

1.5.3 Effect of Extraction Solvent on Carotenoid Recovery during Downstream Processing

The chemical properties of carotenoids must be determined prior to select appropriate methods for their extraction. Appropriate polarity and affinity of the molecule with respect to solvent play a key role in liquid–liquid extraction. For example, phycobilin does not have a phytol chain and is covalently attached to a water-soluble protein. Therefore, it can be easily extracted via pure water. The extraction of chlorophyll requires organic solvents, for example, methanol, ethanol, and acetone. The majority of carotenoids are readily soluble in organic solvents, for example, petroleum, and are also soluble in fats and oils. For extraction of carotenoids from algae, it first is macerated in a suitable solvent like ethanol or acetone. Later, effective cell disruption methods such as ultrasonication, bead milling, and pressure disruption are used to improve cell disruption levels to recover the maximum possible algal pigments. Subsequently, the carotenoid content can be determined spectrometrically and with HPLC. Carotenoids can be recognized by advanced methods such as HPLC-PDA-APCI-IT-TOF-MS, and then algal extracts can be determined for several bioactivities.

Each type of algae needs a suitable extraction method to obtain the highest recovery of pigments. Extraction pretreatment with alcohol such as ethanol and methanol-based saponification with/without acetone reaction, then extraction by hexane:ethanol (1:1, v/v) inhibits the detection of trans-neoxanthin, 9- or 9'-cis-neoxanthin, cis-neoxanthin, antheraxanthin, cis-violaxanthin, chlorophyll b, trans-zeaxanthin, 9-cis-α-carotene, and cis-β-carotene (Soares et al., 2016). However, the preferred method is saponification because it will simplify the later steps, separation and purification, to achieve the ultimate compound with extreme purity. This method leads to deterioration and lack of information on indigenous compounds, and non-saponified extracts show complex chromatograms and higher pigment information, for example, esterified xanthophyll (Soares et al., 2016).

Proper solvent extraction and presoak time had a significant impact on the extraction. The energy-efficient pretreatment for extracting astaxanthin in a short reaction time of 1 minute using the ionic liquid $EtSO_4$ as a green solvent was found promising to get a high astaxanthin yield of 19.5 pg cell^{-1} from *H. pluvialis*. Pre-germination treatment of ionic liquid (IL) was an advanced and effective method for astaxanthin extraction. It showed that there is a great opportunity for a new and effective method for extracting astaxanthin from cysts of *H. pluvialis*. They do not require excessive energy for increased temperatures or intensive powered cell destruction and avoid or reduce the usage of volatile organic solvents.

The difference in efficiency can be seen when using different extraction solvents in previous research. They compared hexane only and hexane/ethanol (1:1) with *C. protothecoides* and *S. dimorphus*. The results indicated that lipid recovery rates were 16–25% higher with both algae in the hexane-only system. In a previous study, *Oedogonium* macroalgae were subjected to cell disruption using hexane as the solvent, which produced the highest chlorophyll yield of 14.02 mg ml^{-1} and the lowest yield of 1.39 mg ml^{-1} in the presence of osmotic shock. These proved that cell disruption methods are substantial in the diffusion of algae pigments in the aqueous state. Tables 1.1 and 1.2 cover the details of extraction methods used for important carotenoids. Lutein extraction from *Chlorella sorokiniana* Kh12 was enhanced by methanol extraction when 8 min of bead milling was adopted (Patel et al., 2022). Another study showed that ultrasound-assisted solvent extraction was more efficient (5.5–10.4 µg g^{-1}) than the saponified solvent extraction method. Maximum zeaxanthin extraction is reported from a chloroform:methanol (1:2) system up to 32.2 mg g^{-1} (Mitra and Mishra, 2019). Pressure-assisted liquid extraction was also effective to recover zeaxanthin from microalgal *Chlorella ellipsoidea* cells (Mitra and Mishra, 2019).

1.6 GLOBAL CAROTENOID MARKET

Due to increasing demand for natural bioactive products, current economic drive, and high fiscal rewards, different microalgae producer industries, especially in the Western hemisphere (Fermentalg, BDIBioLife Science, Allmicroalgae Natural Products, Algalif, Qualitas Health, Triton, Earthrise, A4F, etc.), are emerging as important players in the carotenoid market. The market value of β-carotene is expected to increase over the next few years due to its critical use in the food industry and beverage. The global astaxanthin market value is predicted to be about US$647 million and is likely to grow to US$965 million by 2026, with an 8.3% CAGR. Previously this market was rated at US$555.4 million in 2016 (Rahman, 2020). Latin America, North America, Asia Pacific, Europe, the Middle East, and Africa are the major markets for astaxanthin globally. One of the most important reasons for North American control over the market is due to increasing demand for carotenoids in food supplements, animal feed, food, and healthcare products (Rahman, 2020). With growing economies, South America, China, and India are expected to be large markets in the dietary supplement and pharmaceutical industry over the next six years according to a study of global phycobiliprotein market growth, share, size, trends, and forecasts from 2021. The market value of microalgal carotenoids could increase with a compound annual growth rate of 4% from 2019 to 2025, with a market value of US$452.4 million expected by 2025 as studied by the top ten companies in the algal pigments market in 2022. Synthetic pigments are commonly used in cosmetics, food, pharmaceuticals, and dietary supplements. However, harmless natural pigments from plant and algal sources are emerging due to their side effects. According to the Global Natural Pigment Market Report 2021 (Report by Product Type, Application, and by Region), algal pigments are fascinating due to their scale-up potential, easy operation, and environmentally friendly and sustainable properties (Patel et al., 2022). The size of the European natural pigment market is expected to reach US$2.18 billion by 2028, with a CAGR of 10.1% during the forecast period.

In this situation, the following companies worldwide are earning higher revenues through algae-derived pigments and other products

1. Cyanotech Corporation, Hawai, US
2. Earthrise Nutritionals LLC, California, US
3. Algae Health Sciences, US
4. BlueBioTech Int. GmbH, Germany
5. Bluetec Naturals Co. Ltd, China
6. E.I.D Parry Ltd., Chennai, India
7. Shaivaa Algaetech, India
8. Algatechnologies Ltd, Israel
9. Tianjin Norland Biotech Co. Ltd, Tianjin, China
10. AlgaeCan Biotech Ltd, Canada
11. AstaReal AB, Sweden
12. Sochim International, Italy
13. Chlostanin Nikken Nature Co. Limited, Hong Kong
14. DDW The Color House, US

REFERENCES

Arnal, E., Miranda, M., Almansa, I., Muriach, M., Barcia, J.M., Romero, F.J., Diaz-Llopis, M., Bosch-Morell, F., 2009. Lutein prevents cataract development and progression in diabetic rats. *Graefes Arch. Clin. Exp. Ophthalmol.*, 247(1), 115–120.

Arnal, E., Miranda, M., Barcia, J., Bosch-Morell, F., Romero, F., 2010. Lutein and docosahexaenoic acid prevent cortex lipid peroxidation in streptozotocin-induced diabetic rat cerebral cortex. *Neuroscience*, 166(1), 271–278.

Azeem, M., Iqbal, N., Mir, R.A., Adeel, S., Batool, F., Khan, A.A., Gul, S., 2019. Harnessing natural colorants from algal species for fabric dyeing: a sustainable eco-friendly approach for textile processing. *J. Appl. Phycol.*, 31, 3941–3948.

Barufi, J.B., Figueroa, F.L., Plastino, E.M., 2015. Effects of light quality on reproduction, growth, and pigment content of *Gracilaria birdiae* (Rhodophyta: Gracilariales). *Sci. Mar.*, 79(1), 15–24.

Bhuvaneswari, S., Yogalakshmi, B., Sreeja, S., Anuradha, C.V., 2014. Astaxanthin reduces hepatic endoplasmic reticulum stress and nuclear factor-κB-mediated inflammation in high fructose and high fat diet-fed mice. *Cell Stress Chaperones*, 19(2), 183–191.

Catarina, O., Susana, M., Juliana, P., Salvia, B., Filipa, B.P., Rita, C.A., Oliveira, M., Beatriz, P.P., 2020. Pigments content (chlorophylls, fucoxanthin and phycobiliproteins) of different commercial dried algae. *Separations*, 7(2), 33.

Choi, Y.Y., Patel, A.K., Hong, M.E., Chang, W.S., Sim, S.J., 2019. Microalgae bioenergy carbon capture utilization and storage (BECCS) technology: an emerging sustainable bioprocess for reduced CO_2 emission and biofuel production. *Bioresour. Technol. Rep.*, 7, 100270.

Dewanjee, S., Das, S., Das, A.K., Bhattacharjee, N., Dihingia, A., Dua, T.K., Kalita, J., Manna, P., 2018. Molecular mechanism of diabetic neuropathy and its pharmacotherapeutic targets. *Eur. J. Pharmacol.*, 833, 472–523.

Di Lena, G., Casini, I., Lucarini, M., Lombardi-Boccia, G., 2019. Carotenoid profiling of five microalgae species from large-scale production. *Food Res. Int.*, 120, 810–818.

Ebrahimi Mohammadi, K., Arashrad, F., 2016. Effect of different salinity levels on β-carotene production by *Dunaliella* Sp. Isolates from the Maharlu lake, Iran. *Med. Lab. J.*, 10(5), 58–64.

Hong, M.E., Chang, W.S., Patel, A.K., Oh, M.S., Lee, J.J., Sim, S.J., 2019a. Microalgae-based carbon sequestration by converting LNG-fired waste CO_2 into red gold astaxanthin: the potential applicability. *Energies*, 12, 1718.

Hong, M.E., Yu, B.S., Patel, A.K., Choi, H.I., Song, S., Sung, Y.J., Chang, W.S., Sim, S.J., 2019b. Enhanced biomass and lipid production of *Neochloris oleoabundans* under high light conditions by anisotropic nature of light-splitting $CaCO_3$ crystals. *Bioresour. Technol.*, 287, 121483.

Hu, L.C., 2019. Production of potential coproducts from microalgae. In I-Chen Hu (ed.) *Biofuels from Algae*. Elsevier, 345–358. https://doi.org/10.1016/B978-0-444-64192-2.00014-7

Huang, W., Lin, Y., He, M., Gong, Y., Huang, J., 2018. Induced high-yield production of zeaxanthin, lutein, and β-carotene by a mutant of *Chlorella zofingiensis*. *J. Agric. Food Chem.*, 66, 891–897.

Khan, M.I., Shin, J.H., Kim, J.D., 2018. The promising future of microalgae: current status, challenges, and optimization of a sustainable and renewable industry for biofuels, feed, and other products. *Microb. Cell Fact.*, 17, 36.

Kou, L., Du, M., Zhang, C., Dai, Z., Li, X., Zhang, B., 2017. The hypoglycemic, hypolipidemic, and antidiabetic nephritic activities of zeaxanthin in diet-streptozotocin-induced diabetic Sprague Dawley rats. *Appl. Biochem. Biotechnol.*, 182, 944–955.

Kou, Y., Liu, M., Sun, P., Dong, Z., Liu, J., 2020. High light boosts salinity stress-induced biosynthesis of astaxanthin and lipids in the green alga *Chromochloris zofingiensis*. *Algal Res.*, 50, 101976.

Krinsky, N.I., Johnson, E.J., 2005. Carotenoid actions and their relation to health and disease. *Mol. Asp. Med.*, 26(6), 459–516.

Lamers, P.P., Janssen, M., De Vos, R.C., Bino, R.J., Wijffels, R.H., 2012. Carotenoid and fatty acid metabolism in nitrogen starved *Dunaliella salina*, a unicellular green microalga. *J. Biotechnol.*, 162, 21–27.

Lin, J.H., Lee, D.J., Chang, J.S., 2014. Lutein production from biomass: marigold flowers versus microalgae. *Bioresour. Technol.*, 184, 421–428.

Liu, M., Liu, H., Xie, J., Xu, Q., Pan, C., Wang, J., Wu, X., Zheng, M., Liu, J., 2017. Anti-obesity effects of zeaxanthin on 3T3-L1 preadipocyte and high fat induced obese mice. *Food Funct.*, 8(9), 3327–3338.

Lopez, M.J., Hall, C.A., 2021. *Physiology, Osmosis*. StatPearls Publishing. www.ncbi.nlm.nih.gov/books/NBK557609/.

Maoka, T., 2020. Carotenoids as natural functional pigments. *J. Nat. Med.*, 74(1), 1–16.

Martinez-Andrade, K.A., Lauritano, C., Romano, G., Ianora, A., 2018. Marine microalgae with anti-cancer properties. *Mar. Drug.*, 16(5), 165.

Mitra, M., Mishra, S., 2019. A comparative analysis of different extraction solvent systems on the extractability of eicosapentaenoic acid from the marine eustigmatophyte *Nannochloropsis oceanica*. *Algal Res.*, 38, 101387.

Mrowicka, M., Mrowicki, J., Kucharska, E., Majsterek, I., 2022. Lutein and zeaxanthin and their roles in age-related macular degeneration-neurodegenerative disease. *Nutrients*, 14(4), 827.

Muriach, M., Bosch-Morell, F., Alexander, G., Blomhoff, R., Barcia, J., Arnal, E., Almansa, I., Romero, F.J., Miranda, M., 2006. Lutein effect on retina and hippocampus of diabetic mice. *Free Radic. Biol. Med.*, 41(6), 979–984.

Muriach, M., Bosch-Morell, F., Arnal, E., Alexander, G., Blomhoff, R., Romero, F., 2008. Lutein prevents the effect of high glucose levels on immune system cells in vivo and in vitro. *J. Physiol. Biochem.*, 64(2), 149–157.

Novosel, N., Mišić Radić, T., Zemla, J., Lekka, M., Čačković, A., Kasum, D., Legović, T., Žutinić, P., Gligora Udovič, M., Ivošević DeNardis, N., 2022. Temperature-induced response in algal cell surface properties and behaviour: an experimental approach. *J. Appl. Phycol.*, 34(1), 243–259.

Park, C.H., Xu, F.H., Roh, S.-S., Song, Y.O., Uebaba, K., Noh, J.S., Yokozawa, T., 2015. Astaxanthin and Corni Fructus protect against diabetes-induced oxidative stress, inflammation, and advanced glycation end product in livers of streptozotocin-induced diabetic rats. *J. Med. Food*, 18(3), 337–344.

Patel, A.K., Choi, Y.Y., Sim, S.J., 2020a. Emerging prospects of mixotrophic microalgae: way forward to bioprocess sustainability, environmental remediation and cost-effective biofuels. *Bioresour. Technol.*, 300, 122741.

Patel, A.K., John, J., Hong, M.E., Sim, S.J., 2019. Effect of light conditions on mixotrophic cultivation of green microalgae. *Bioresour. Technol.*, 282, 245–253.

Patel, A.K., John, J., Hong, M.E., Sim, S.J., 2020b. A sustainable mixotrophic microalgae cultivation from dairy wastes for carbon credit, bioremediation and lucrative biofuels. *Bioresour. Technol.*, 313, 123681.

Patel, A.K., Laroche, C., Marcati, A., Violeta, A.U., Jubeau, S., Marchal, L., Petit, E., Djelveh, G., Michaud, P., 2013. Separation and fractionation of exopolysaccharide from *Porphyridium cruentum*. *Bioresour. Technol.*, 145, 345–350.

Patel, A.K., Singhania, R.R., Awasthi, M., Varjani, S., Bhatia, S.K., Tsai, M.L., Hseih, S.L., Chen, C.W., Dong, C.D., 2021c. Emerging role of macro- and microalgae as prebiotic. *Microb. Cell. Fact.*, 20, 112.

Patel, A.K., Singhania, R.R., Chang, J.S., Chen, C.W., Dong, C.D., 2021b. Novel application of biodesalination from microalgae. *Bioresour. Technol.*, 337, 125343.

Patel, A.K., Singhania, R.R., Chen, C.W., Dong, C.D., 2021f. Algal polysaccharide: current status and future perspectives. *Phytochem. Rev.* https://doi.org/10.1007/s11101-021-09799-5.

Patel, A.K., Singhania, R.R., Dong, C.D., Obulisami, P.K., Sim, S.J., 2021e. Mixotrophic biorefinery: a promising algal platform for sustainable biofuels and high value coproducts. *Renew. Sust. Energ. Rev.*, 152, 111669.

Patel, A.K., Singhania, R.R., Sim, S.J., Dong, C.D., 2021a. Recent advancements in mixotrophic bioprocessing for production of high value microalgal products. *Bioresour. Technol.*, 320, 124421.

Patel, A.K., Singhania, R.R., Wu, C.H., Kuo, C.H., Chen, C.W., Dong, C.D., 2021d. Advances in micro- and nanobubbles technology for application in biochemical processes. *Environ. Technol. Innov.*, 23, 101729.

Patel, A.K., Vadrale, A.P., Tseng, Y.S., Chen, C.W., Dong, C.D., Singhania, R.R., 2022. Bioprospecting of marine microalgae from Kaohsiung seacoast for lutein and lipid production. *Bioresour. Technol.*, 351, 126928.

Perales-Vela, H.V., Salcedo-Álvarez, M.O., Parra-Marcelo, R., Gaviria-González, L.C., de Jesús Martínez-Roldán, A., 2021. Growth and metabolic responses to methyl viologen (1, 1′-dimethyl-4, 4′-bipyridinium dichloride) on *Chlorella vulgaris*. *Chemosphere*, 281, 130750.

Pereira, A.G., Otero, P., Echave, J., Carreira-Casais, A., Chamorro, F., Collazo, N., Jaboui, A., Lourenço-Lopes, C., Simal-Gandara, J., Prieto, M.A., 2021. Xanthophylls from the sea: algae as source of bioactive carotenoids. *Mar. Drug.*, 19(4), 188.

Perez-Garcia, O., Bashan, Y., 2015. Microalgal heterotrophic and mixotrophic culturing for bio-refining: from metabolic routes to techno-economics. In A. Prokop (ed.), *Algal biorefineries*. Springer International Publishing. https://doi.org/10.1007/978-3-319-20200-6-3.

Piotrowska-Niczyporuk, A., Bajguz, A., Zambrzycka, E., Godlewska-Żyłkiewicz, B., 2012. Phytohormones as regulators of heavy metal biosorption and toxicity in green alga *Chlorella vulgaris* (Chlorophyceae). *Plant Physiol. Biochem.*, 52, 52–65.

Purbonegoro, T., Puspitasari, R., Suratno, S., Aji, A.S., 2018. Toxicity of copper (Cu) on the growth and chlorophyll-a contents of marine microalgae *Isochrysis sp. AIP Conf. Proc.*, 2026, 020007.

Rahman, K., 2020. Food and high value products from microalgae: market opportunities and challenges. In *Microalgae Biotechnology for Food, Health and High Value Products*, 3–27. https://doi.org/10.1007/978-981-15-0169-2_1.

Rajput, A., Singh, D.P., Khattar, J.S., Swatch, G.K., Singh, Y., 2021. Evaluation of growth and carotenoid production by a green microalga *Scenedesmus quadricauda* PUMCC 4.1.40. under optimized culture conditions. *J. Basic Microbiol.*, 62, 1156–1166. https://doi.org/10.1002/jobm.202100285.

Ren, Y., Deng, J., Huang, J., Wu, Z., Yi, L., Bi, Y., Chen, F., 2021a. Using green alga *Haematococcus pluvialis* for astaxanthin and lipid co-production: advances and outlook. *Bioresour. Technol.*, 340, 125736.

Ren, Y., Sun, H., Deng, J., Huang, J., Chen, F., 2021b. Carotenoid production from microalgae: biosynthesis, salinity responses and novel biotechnologies. *Mar. Drug.*, 19(12), 713.

Sánchez, J.F., Fernández, J.M., Acién, F.G., Rueda, A., Pérez-Parra, J., Molina, E., 2008. Influence of culture conditions on the productivity and lutein content of the new strain *Scenedesmus almeriensis*. *Proc. Biochem.*, 43, 398–405.

Sarnaik, A., Nambissan, V., Pandit, R., Lali, A., 2018. Recombinant *Synechococcus elongatus PCC 7942* for improved zeaxanthin production under natural light conditions. *Algal Res.*, 36, 139–151.

Sharma, A., Soulange, J., Driver, M.F., Ambati, R.R., Gokare, R., Neetoo, H., 2019. *Handbook of Algal Technologies and Phytochemicals: Volume-I: Food, Health and Nutraceutical Applications*. CRC Press.

Sharma, D., Bhattacharya, P., Kalia, K., Tiwari, V., 2017. Diabetic nephropathy: new insights into established therapeutic paradigms and novel molecular targets. *Diabetes Res. Clin. Pract.*, 128, 91–108.

Sim, S.J., John, J., Hong, M.E., Patel, A.K., 2019. Split mixotrophy: a novel mixotrophic cultivation strategy to improve mixotrophic effects in microalgae cultivation. *Bioresour. Technol.*, 291, 121820.

Soares, A.T., Marques Júnior, J.G., Lopes, R.G., Derner, R.B., Antoniosi Filho, N.R., 2016. Improvement of the extraction process for high commercial value pigments from *Desmodesmus sp. Microalgae. Braz. Chem. Soc.*, 27(6), 1083–1093.

Sun, Z., Zhang, Y., Sun, L.P., Liu, J., 2019. Light elicits astaxanthin biosynthesis and accumulation in the fermented ultrahigh-density *Chlorella zofinginesis*. *J. Agric. Food Chem.*, 67(19), 5579–5586.

Sung, Y.J., Patel, A.K., Yu, B.S., Kim, J., Choi, H.I., Sim, S.J., 2019. Sedimentation rate-based screening of oleaginous microalgal for fuel production. *Bioresour. Technol.*, 293, 122045.

Wang, L., Liu, Z., Jiang, H., Mao, X., 2021. Biotechnology advances in β-carotene production by microorganisms. *Trends Food Sci. Technol.*, 111, 322–332.

Wang, X., Zhang, M.M., Sun, Z., Liu, S.F., Qin, Z.H., Mou, J., Zhou, Z.G., Lin, C.S.K., 2020b. Sustainable lipid and lutein production from Chlorella mixotrophic fermentation by food waste hydrolysate. *J. Hazard. Mater.*, 400, 123258.

Wang, X., Zhang, P., Wu, Y., Zhang, L., 2020a. Effect of light quality on growth, ultrastructure, pigments, and membrane lipids of *Pyropia haitanensis*. *J. Appl. Phycol.*, 32(6), 4189–4197.

Xu, L., Zhu, J., Yin, W., Ding, X., 2015. Astaxanthin improves cognitive deficits from oxidative stress, nitric oxide synthase and inflammation through upregulation of PI3K/Akt in diabetes rat. *Int. J. Clin. Exp. Pathol.*, 8(6), 6083.

Yeh, P.T., Huang, H.W., Yang, C.M., Yang, W.S., Yang, C.H., 2016. Astaxanthin inhibits expression of retinal oxidative stress and inflammatory mediators in streptozotocin-induced diabetic rats. *PLoS One*, 11(1). https://doi.org/10.1371/journal.pone.0146438.

2 Innovative Extraction Methods to Obtain Bioactive Compounds from Aquatic Biomass

Ana Lucía Sarmiento-Padilla, Blanca E. Morales-Contrerasa, Rohit Saxena, Ruth Belmares, Adriana M. Bonilla Loaiza, K.D. González-Gloria, Araceli Loredo-Treviño, Rosa M. Rodríguez-Jasso and Héctor A. Ruiz

CONTENTS

2.1	Introduction	21
2.2	Macroalgae (Seaweeds)	22
2.3	Microwave	23
2.4	Ultrasound	24
2.5	Pulsed Electric Field Treatment	25
2.6	Sub- and Supercritical Fluids	26
2.7	Hydrothermal Processing	27
2.8	Others	28
2.9	Perspectives	29
2.10	Conclusions	29
2.11	Acknowledgments	29
	References	30

2.1 INTRODUCTION

Macro and micro algae are photosynthetic organisms capable of fixing carbon dioxide (CO_2). They are aquatic plants composed of several molecules of interest which are fermentable carbohydrates. These carbohydrates can be used in the generation of biofuels and other products of cosmetic, industrial and food interest (Figure 2.1). The recovery or release of these carbohydrates is always a challenge, and in this chapter, some strategies are discussed. Macroalgae are also known as seaweeds and are divided into red, green and brown algae according to the type of photosynthetic pigment they contain, so they are *Chlorophyta* (green algae), *Rhodophyta* (red algae) and *Phaeophyta* (Brown algae) [1]. Their composition also varies. They are marine life. Microalgae, on the other hand, are organisms with a cell size between 2 and 200 μm and a great capacity to fix CO_2. They can be autotrophic or heterotrophic and are capable of generating a large amount of biomass that will contain lipids, proteins and/or carbohydrates, depending on the species and culture conditions, just like macroalgae. The classification of these organisms follows both morphological and genetic differences. The most general classification divides them into *Chlorophyta* (fresh and salt water and some terrestrial environments, with chlorophyll a and b; they are green); *Rhodophyta* (mainly salt water with phycocyanin and phycoerythrin; they are olive-brown), *Haptophyta* (mainly marine, with

DOI: 10.1201/9781003326946-2

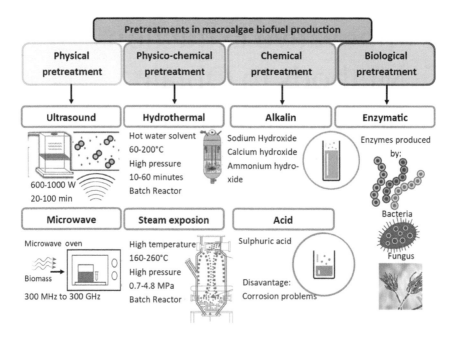

FIGURE 2.1 Process of algae biomass for biofuel production and high added value compounds.

one or two chloroplasts with pyrenoids; yellowish-green color), *Stramenopiles* (two groups: *Eustigmatophyceae* and *Bacillariophiceae*, with chlorophyll a and c, respectively) and *Dinophyta* (mostly marine life, heterotrophic). A great advantage of algae is that they do not compete with terrestrial plants for space [2]. The objective of this chapter is to present the different types of innovative methods for obtaining high value added compounds from macroalgae and microalgae, as well as to analyze the principles, advantages and disadvantages of each of them. Macroalgae and microalgae are biomasses with great potential as sources for obtaining different types of compounds due to their great availability and unique composition; however, depending on the type of macro and micro algae and the compound to be extracted, different extraction methods can be used.

2.2 MACROALGAE (SEAWEEDS)

The macroalgae (or seaweeds) are a promising source of biomaterials, which can have multiple applications in various sectors due to their composition and the way in which they are produced, since they do not need large territories and fresh water for their production, so it is possible to obtain high added value products [3].

Macroalgae can be a source of compounds such as sulfated galactans; glucuronoxylorhamnans; glucuronoxylorhamnogalactans; xyloarabinogalactans; alginates; uronide polymers composed of mannuronate and guluronate; laminarin; agar; fucoidan; monosaccharides such as fucose, mannose, xylose and glucose; carrageenan; antioxidants; carotenoids; phenols; fucoxanthin; protein; lipids; and minerals, just to mention a few [1]. All of these compounds have different applications as thickeners, stabilizers, gelling agents and emulsifiers in the cosmetic and pharmaceutical industries, and some can be uses as precursors in the production of biofuels. In addition, macroalgae are becoming an essential product in aquaculture, as they do not compete with other crops for land and freshwater [4].

Traditional methods such as maceration and high temperatures have been used for centuries to obtain compounds of interest from macroalgae, these involve stages such as the collection of macroalgae, cleaning and drying. However, due to the current situation in terms of the shortage of renewable sources of materials, as well as the increase in the population and the demand for food, different ways of obtaining these compounds have emerged, and the generation of new alternatives has been necessary. Important points must be taken into account when it comes to the separation of macroalgae compounds: even more so when it comes to bioactives, the development of precise, cost-effective, fast and environmentally friendly extraction processes must be satisfied.

2.3 MICROWAVE

The use of microwave assisted extraction (MAE) for compound extraction dates back to 1986. This technology is based on converting microwaves into heat through the mechanisms of dipole rotation and ionic conduction. This heat has the ability to transfer energy to solvents, which causes the migration of dissolved ions due to the formation of interruptions in the hydrogen bonds and, as a consequence, a greater penetration of the solvent in the matrix (Figure 2.2 shows the general extraction process using MAE); there are also variants of MAE, such as the variant without solvent (SFME) or others such as pressure, atmosphere variations (by using noble gases) or with reflux; it has been considered a reliable green technology for the extraction of high added value compounds [5].

The technique can be used in different types of systems, such as open or closed containers. In the first case, extractions are carried out at low pressures and temperatures, while in the second case, compounds are extracted at higher temperatures and pressures; commonly the time used for extraction with this method is from 1 to 40 minutes.

Some of its advantages are the use of water as a solvent, a lower amount of waste and the speed of the process compared to others such as extraction processes by maceration or Soxhlet, reducing times of up to 24 h by around 30 minutes. It is a process that can be easily carried out at an industrial level by making extraction possible in less time and with higher yield than conventional methods. This technology also allows the reduction of use of solvents, which as a consequence also brings higher performance in the extraction of compounds of interest by being in contact with high temperatures or solvents for a shorter amount of time; However, the temperature must be carefully monitored, since not only could heat cause boiling of the solvent used, but it is also possible to cause degradation of the compound of interest; it is also necessary to carry out a process of separation of solids from solvent [6].

The extracted compounds are usually lipophilic in nature; some of the compounds extracted from algae using this technology are lipids, essential oils, carotenoids, fatty acids, polysaccharides, hybrid carrageenans and antioxidants [7].

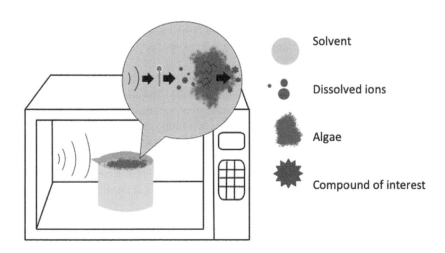

FIGURE 2.2 Schematization of the extraction process by means of MAE.

Perspectives: Despite being a technology that has already been investigated, it is a subject that still needs a lot to know and experience, as well as processes already explored to optimize. It is likely that through study, the parameters to be used to make MAE a suitable process for the extraction of hydrophilic compounds will be found.

Brief conclusion: Microwave-assisted extraction is an increasingly used technology due to the large amount of research that has been carried out so far, and it will continue to increase due to advantages such as great energy and economic savings, as well as the preservation of the environment.

2.4 ULTRASOUND

Ultrasound is an ultrasonic wave generated by a vibrating source with a threshold higher than that of the human ear (frequencies > 20 kHz), which causes the transfer of propagation energy (through the medium) known as ultrasound power (W), ultrasound intensity (W/cm^2) or acoustic energy density (W/cm^3 or W/mL); formation of cavitation bubbles and subsequently the transfer of energy to the surrounding particles [8].

By means of ultrasound, it is possible to break the algae cell wall through the phenomena of cavitation, vibration and crushing, resulting in mechanical and thermal effects and thus achieving the extraction of compounds of interest. The cavitation phenomenon occurs when the negative pressure reaches a critical value in which the liquid can form a cavity of gas or vapor in the environment that is transported by the medium to the raw material, which is known as the cavitation threshold. This pressure propagates through the medium by means of the ultrasonic wave, alternately expanding and contracting in the liquid, which is called rarefaction and compression, and varying according to the frequency of the pulse of the sound wave, also with high and low pressure areas. The bubbles grow until they reach a critical point, at which they explode, generating temperatures of up to 5000K in the surrounding medium and pressures of up to 100 MPa. As a consequence, they cause turbulence and shear forces in the surrounding medium, which favors the rupture of the cell wall and, consequently, obtain the compounds of interest [9].

Ultrasound is divided into a low-power and high-frequency range and high-power and low-frequency range, the first using frequencies >100 kHz applied with the purpose of non-invasive analysis and the second frequencies between 16 and 100 kHz used for food processing, control and safety [9].

There are two types of ultrasound equipment. One of them causes indirect sonication by means of an ultrasonic bath, while the other causes a direct sonication by means of an ultrasonic probe. In the first type, frequencies between 40 and 50 kHz can be used with powers of 50 to 500 W, and the sample is immersed in the medium, while the second only works at a frequency of 20 kHz, and an ultrasonic probe is inserted into the sample. Figure 2.3 shows the differences between an ultrasonic bath and an ultrasonic probe.

Among its advantages are the low investment costs compared to other extraction techniques; it can be used with a wide variety of solvents (including water) and a low amount of them; high temperatures are not required, so it prevents the degradation of thermolabile compounds; and it reduces extraction time and obtains a higher yield both on a laboratory and industrial scale. However, it is important to know that it can cause degradation in the component of interest if the sonication energy is higher than needed.

This technology has been used for the extraction of high-value compounds such as taurine, polyphenols, fucose, uronic acid, laminarin, phycobiliproteins, total phenols, antioxidants, prebiotic compounds, polysaccharides and carbohydrates [10].

Perspectives: One of the best options for the extraction of compounds that degrade at low temperatures is ultrasound-assisted extraction, so it is certain that work and research will continue on the extraction of this type of compounds, which will generate great advances

Innovative Extraction Methods to Obtain Bioactive Compounds 25

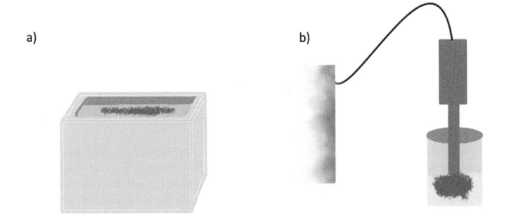

FIGURE 2.3 a) Ultrasonic bath equipment, b) ultrasonic probe equipment.

in the commercialization of them, since most of the current technologies require heat for their function, which in addition to raising the cost can decrease the quantity and quality of the objective compound obtained.

Brief conclusion: Ultrasound is an excellent alternative for the extraction of high added value compounds in macroalgae, as it is an option in which temperature is not required, and the extraction of thermolabile and sensitive compounds is possible without the heat factor and with superior performance.

2.5 PULSED ELECTRIC FIELD TREATMENT

Pulsed electric field treatment (PEF) is based on the principles of the formation of a potential difference across a cell membrane that acts as a capacitor with a low dielectric constant between two electrodes applying short pulses of electricity (μs to ms) with moderate voltage (approximately between 0.5 and 20 kV/cm), which increases the transmembrane potential due to the accumulation of charges across the membrane; subsequently the membrane thins and breaks if the critical breakdown voltage is reached, and the electric field created depends on the applied voltage, the shape of the electrodes and the space between them. Depending on the process parameters, a reversible or irreversible effect can be induced, irreversible being used for the extraction of high added value compounds, while reversible is used for the introduction of specific molecules to the raw material [11].

Its principle is based on the transient increase in the permeability of the plasma membrane due to the application of an electric field (electroporation), and thanks to the permeability obtained, the plasma membrane allows the entry of ions and molecules to the cytoplasm, as well as cell fusion, the insertion of proteins into the membrane or the destruction of the membrane; It shows the extraction process by means of a pulsed electric field [12].

It is a non-thermal technology, which avoids the change of color or flavor and the degradation of nutrients or compounds of interest, and it does not use chemicals and has lower energy requirements than methods such as dehydration. It is used in the food industry for disinfection, modification of enzymatic activity and extraction of high added value compounds. However, a great disadvantage is the limited use that it has in macroalgae due to its high salt content (high conductivity of the media), which means that high currents are required in the electric fields for the permeabilization of the cell membrane, which subsequently also leads to heating that exceeds the energy dissipation capacity of the equipment [12].

By means of PEF, the extraction of polyphenolic compounds, polysaccharides, proteins, isothiocyanates, phytosterols, carbohydrates and lipids has been achieved; extraction of secondary metabolites has been achieved with low treatment intensities.

> **Perspectives**: Despite being a technology that offers many benefits, its use is actually very limited in the specific area of aquatic biomass; probably in the near future there will be an alternative for using it with algae, which would bring great advantages, such as the possible extraction of compounds of interest that are sensitive to certain types of chemicals or solvents.
>
> **Brief conclusion**: This technique has allowed higher yields in shorter extraction times than those obtained with traditional treatments, in addition to being free of chemicals or solvents harmful to health; however, the knowledge for its application is quite narrow, especially for the extraction of water-soluble components or mixtures without subsequent separation and purification, which unfortunately has been solved with the use of solvents to obtain a higher yield of compound of interest according to its polarity, which increases the risk of product contamination.

2.6 SUB- AND SUPERCRITICAL FLUIDS

The diversity in extraction methods to obtain compounds of interest lies in their nature; in the macroalgae compounds case, the vast majority correspond to bioactive compounds [13], as already mentioned, which are generally sensitive to factors such as high temperature, exposure to light and oxidation. The use of aggressive or severe extraction methods is not the best option. Some alternatives have been developed over time and with technology and knowledge advances, like use of sub- and supercritical fluids.

The main advantage of sub- and supercritical fluids as an extraction method is efficient diffusion, which leads to a faster process. The use of supercritical fluids for the extraction of compounds of interest consists of using a fluid under conditions above its temperature and critical pressure [14], and in these conditions, gases and liquids can coexist, which is interesting from the extraction point of view, since they have the ability to diffuse into solids such as a gas, as well as dissolve them as liquids, penetrating different types of matrices, reaching a density similar to a liquid and viscosity similar to a gas, improving mass transfer. On the other hand, subcritical fluids refer to those that are at a temperature above their boiling point but below their critical temperature under a pressure that keeps the fluid in its liquid state.

The main advantage of the use of sub- and supercritical fluids as alternative method to obtain macroalgal compounds in comparison to others is the reduction of solvent consumption and extraction times [14]. On the other hand, sub-critical water extraction is considered a zero toxic waste method because water is used as an extraction solvent, contrary to the typical solvents such as hexane, methanol, chloroform and ethanol, to name a few.

> **Perspectives:** The use of sub- and supercritical fluids is a promising technology for obtaining compounds of interest from macroalgae, since it has been proven that high extraction yields are obtained in relatively short periods of time. However, it is necessary to delve into the conditions of the use of this method in different types of biomass, as well as the optimization of parameters depending on the type of compound of interest, since in some cases it is necessary to apply pretreatments with other technologies.
>
> **Brief conclusion:** This type of technology will become more popular in coming years with the improvement of equipment for the application of sub- and supercritical fluids, since, under the operating conditions mentioned in this section, the improvements in the transfer of mass in the processes are evident due to the properties that fluids take on both above and below critical conditions, making these an effective and competitive extraction method

compared to traditional methods since they are methodologies that considerably reduce the use of organic solvents and energy consumption.

2.7 HYDROTHERMAL PROCESSING

Hydrothermal processing as a method for obtaining high added value compounds refers to the breakdown of the cell wall and release of different compounds into the liquid phase by water action at high temperatures and pressures [3]. The source of energy to achieve high temperatures and pressures causes variations in the application of this principle to obtain compounds of interest. Some of the most used are steam explosion and microwaves.

Obtaining different compounds of interest from macroalgae is based mainly on modifying the cellular structure of biomass, which allows subsequent extraction; for this, in recent years, innovative pretreatments such as the compressional-puffing-hydrothermal process (which consists of a first heating stage [140–220°C] under atmospheric pressure conditions and a second stage in which the pressure is reduced with superheated water) have been proposed, which facilitates the extraction of compounds like fucoidan [15].

Hydrothermal treatment is considered a clean method due to the advantages it represents. For the specific application in macroalgae, it has been used in dry and wet algae. The disadvantage of applying it to dry biomass is the energy expenditure necessary to be able to dry them, but even so it is still considered viable. On the other hand, in wet biomass, hydrothermal treatments can be divided into three types: 1) hydrothermal gasification (HTG), 2) hydrothermal carbonization and 3) hydrothermal liquefaction (HTL) [4].

In the case of HTG, although it is not considered an extraction method, it is considered a conversion method, obtaining products of interest such as hydrogen-enriched gas, with great potential from the energy point of view. In this sense, the production of biofuels has seen a boom in recent years due to the global environmental problems that we are facing. Bio-hydrogen is considered a clean and sustainable source of energy because the absence of carbon, and a good method to obtain it is the application of HTG. The advantage of applying it to wet macroalgae is that it acts as reaction medium and reactant. In addition, the use of tools such as mathematical modeling to contribute to a better understanding of the conversion process is being explored, since it is essential to optimize the processes, understand the kinetics involved and improve scaling not only to improve them but also to be able to adapt them to biomass variability.

With respect to HTC, it is a technique considered environmentally friendly with great potential in the production of high-energy density solid fuels and to produce value-added products. Parameters such as temperature, residence time and pressure are essential to achieve effectiveness in different types of biomass, including macroalgae [16]. HTC offers the advantage of using the water already present in the macroalgae as a solvent and catalyst for the conversion reactions. Macroalgae such as *sargassum* have received particular interest for this type of exploitation due to their abundance and availability [16]. HTC can dehydrate, decarboxylate and deaminate different types of biomass to produce hydrochar with similar characteristics to those of coal. As a perspective in the future of the application of this type of methodology, it is aimed at studying the effect of the biochemical composition of biomass on the properties of the hydrocarbons obtained to later diversify their possible applications [17].

HTL is the thermochemical transformation of biomass into liquid fuels by processing the solid biopolymeric structure into liquid components using a pressurized hot water environment for a sufficient period of time. HTL has been studied in various types of biomass to obtain biofuels, showing encouraging results for the particular case of macroalgae that are generally composed of lipids, soluble polysaccharides, proteins and significant amounts of sulfur and nitrogen. It favors the reduction in the necessary energy to achieve the conversion of biomass in comparison to that which comes from terrestrial plants, which by lacking the latter significantly increases the energy expenditure. The hydrochar produced from macroalgae has a calorific value similar to that of a low-range coal, as well as significant demineralization compared to the original macroalgae, suggesting its safe use as a solid combustion fuel.

> **Perspectives:** Due to the great diversification in the application of hydrothermal processes, it is necessary to continue with the exploration of alternatives for the optimization of the processes, the exploitation of macroalgae as biomass to obtain high added value compounds and the obtaining of biofuels in combination with this technology shows great potential that will reduce energy costs by proposing different alternatives that are more environmentally friendly than many of those currently being exploited.
>
> **Brief conclusion:** Taking into account the current advances in hydrothermal processes, these can be considered emerging technologies. When used as pretreatments, they are crucial in the fractionation of the compounds present in macroalgae, with possible exploitation. Hydrothermal treatments are commonly combined with other technologies, as already described in this section, mainly to supply the energy necessary to achieve the appropriate conditions for obtaining the final products, so research will continue to diversify its potential in this regard.

2.8 OTHERS

In recent years, other alternatives have emerged to obtain compounds of interest and take advantage of the biomass of the different types of macroalgae. One of these methods is the enzymatic extraction method. In general, it has presented many advantages due to its great specificity; however, the production of enzymes remains a great challenge for industries. Due to the specificity of the enzymes, the investigations in this sense are the same. For example, in the particular case of brown algae, a study has been done on the application of the enzyme laminarinase, capable of hydrolyzing the compound laminarin, present in a percentage of around 35%. This compound is characterized by having a D-mannitol residue (M-chains), while others terminate with a reducing 3-linked glucose residue (G-chains) [18]. The use of hydrolytic enzymes in the exploitation of macroalgae under the biorefinery concept after obtaining compounds of great interest such as agar has allowed the release of other compounds linked to the macroalgae cell wall, such as phenolic compounds, as well as the release of the polysaccharide fraction as monomers and oligomers, including hydrolysis of the protein fraction more effectively.

Another innovative method is what is known as reactive extrusion, which is a method that combines the use of extrusion with a chemical process such as the synthesis or modification of polymers that can be transferred to the extraction of macroalgae. On the other hand, there is a method called photoblanching, which is based on absorption and photochemical reactions generated by solar radiation, degrading the coloration of the organic material, allowing the transfer of energy through the transfer of electrons and free radicals in the biomass. The effectiveness of the method has been proven in agar extraction, showing better quality compared to other methods in addition to being an ecofriendly alternative [19].

> **Perspectives:** Technology is advancing by leaps and bounds, as well as the generation of new information that has allowed the development of new alternatives; even the combination of different methodologies can offer greater advantages. Since currently it is not enough to design methods that achieve good yields – they also have to be environmentally friendly and economical—new and better alternatives will continue to emerge in the coming years, and the use of macroalgae as a source of biomass to obtain compounds will continue to grow.
>
> **Brief conclusion:** Enzymatic methods are an excellent option for obtaining high added value compounds from macroalgae because they have excellent yields, and they are highly specific and friendly to the environment; However, it is necessary to improve the methods of

obtaining enzymes to scale the processes to the industrial level. On the other hand, the use of methods like reactive extrusion and photoblanching sounds promising, but more research will be necessary to be able to reach large-scale application.

2.9 PERSPECTIVES

Both macro and microalgae are a good renewable source of food and a source of substrate for the generation of fuels and molecules of importance for different industries, as discussed.

The extraction of carbohydrates from algal biomass is essential for the production of biofuels, that is, the pretreatment of the biomass. As discussed in this chapter, there are different methodologies to achieve the rupture of the cell walls of algae biomass. Likewise, it is important to consider the size of the cells and if the organism is multi-celled or unicellular, as well as its composition.

Different techniques for biomass pretreatment will have advantages and disadvantages. There are treatments that will achieve a good cellular breakdown but can also degrade polysaccharides. Others may be more gentle, but they will not be as successful in breaking down the cell wall.

Various aspects must be taken into account when choosing the pretreatment, in addition to the type of algae. In general, these factors are the cost of the infrastructure, the operating cost, the impact on the environment and, above all, the yields and degradation state of fermentable carbohydrates for the production of biofuels and molecules of importance.

Hydrothermal methods are simple methods with little environmental impact. They achieve good substrate recovery, and there is little wear and tear on the equipment. The operation is straightforward, but good equipment must be invested in to achieve these results. The same goes for the microwave and electric pulse methods; however, the equipment needs to be more specialized.

Due to the growing interest in algae, both for their molecules with biological properties, their importance as a substrate for obtaining biofuels and the problems of their overpopulation, techniques for their pretreatment and subsequent bioconversion should continue to be studied, and efforts should be made to scale them up to meet the demand for less polluting fuel as well as products that have benefits for health and different industries.

2.10 CONCLUSIONS

It is clear that extraction methods evolve more every day, and all the technologies detailed in this chapter have been shown to be more efficient and effective than traditional extraction methods, shortening extraction times as well as bringing higher yields of the compounds of interest, preserving their viability, as well as caring for the environment; however, traditional extraction techniques are still used to complement the new ones to obtain better results, one of the ones used being maceration (which does not generate an environmental impact). On the other hand, the use of organic solvents continues, even with innovative extraction technologies to improve the results obtained through the affinity that a solvent can have with the compound of interest, which has a negative impact on the environment, as well as putting in doubt the viability of the product obtained for human consumption. This reflects a clear opportunity for improvement and research to find alternative ways of increasing yield or facilitating the extraction of a specific compound of interest without having to resort to organic solvents, especially before implementing them at an industrial level, as well as clearly defining the mechanisms of action of new technologies for the extraction of high added value compounds. It is a challenge that we are facing, but thanks to technological progress, there are increasing opportunities to meet it.

2.11 ACKNOWLEDGMENTS

We gratefully acknowledge support for this research by the Mexican Science and Technology Council (CONACYT, Mexico) for the Infrastructure Project—FOP02–2021–04 (Ref. 317250) and the

Innovation Incentive Program (PEI)—Mexican Science and Technology Council (SEP-CONACYT) with the Project (Ref. PEI-251186). The authors Ana Lucía Sarmiento-Padilla, Rohit Saxena, Adriana M. Bonilla Loaiza and K.D. González-Gloria thank the Mexican National Council for Science and Technology (CONACYT) for their PhD Fellowship (grant number: 1004500, 1013150, 750836, 785884, respectively), and Dr. Blanca E. Morales-Contrerasa thanks the Mexican Science and Technology Council (CONACYT) for her post-doctoral fellowship (grant number: 479743).

REFERENCES

[1] A. Lara, R.M. Rodríguez-Jasso, A. Loredo-Treviño, C.N. Aguilar, A.S. Meyer, and H.A. Ruiz, *Enzymes in the Third Generation Biorefinery for Macroalgae Biomass*, 2020.

[2] J. Velazquez-Lucio, et al., Microalgal biomass pretreatment for bioethanol production: A review. *Biofuel Res. J.* 5, no. 1 (2018), pp. 780–791. doi: 10.18331/BRJ2018.5.1.5.

[3] D.E. Cervates-Cisneros, D. Arguello-Esparza, A. Cabello-Galindo, B. Picazo, C.N. Aguilar, H.A. Ruiz, et al., Hydrothermal processes for extraction of macroalgae high value-added compounds, in *Hydrothermal Processing in Biorefineries: Production of Bioethanol and High Added-Value Compounds of Second and Third Generation Biomass*, H.A. Ruiz, T.M. Hedegaard and H. Trajano, eds., Cham: Springer, 2017, pp. 461–481.

[4] K. Sudhakar, R. Mamat, M. Samykano, W.H. Azmi, W.F.W. Ishak, and T. Yusaf, An overview of marine macroalgae as bioresource. *Renew. Sustain. Energy Rev.* 91 (2018), pp. 165–179.

[5] R.G. Araújo, R.M. Rodriguez-Jasso, H.A. Ruiz, M. Govea-Salas, M.E. Pintado, and C.N. Aguilar, Process optimization of microwave-assisted extraction of bioactive molecules from avocado seeds. *Ind. Crops Prod.* 154 (2020), p. 112623.

[6] A. de S. e. Silva, W.T. de Magalhães, L.M. Moreira, M.V.P. Rocha, and A.K.P. Bastos, Microwave-assisted extraction of polysaccharides from *Arthrospira* (Spirulina) *platensis* using the concept of green chemistry, *Algal Res.* 35 (2018).

[7] C. Wen, J. Zhang, H. Zhang, C.S. Dzah, M. Zandile, Y. Duan, et al., Advances in ultrasound assisted extraction of bioactive compounds from cash crops—a review. *Ultrason. Sonochem.* 48 (2018), pp. 538–549.

[8] I. Lavilla and C. Bendicho, *Fundamentals of Ultrasound-Assisted Extraction*, Elsevier Inc., 2017.

[9] A.M. Ciko, S. Jokić, D. Šubarić and I. Jerković, Overview on the application of modern methods for the extraction of bioactive compounds from marine macroalgae, *Mar. Drugs* 16 (2018).

[10] N. Flórez-Fernández and M.J. González Muñoz, Ultrasound-assisted extraction of bioactive carbohydrates, in *Water Extraction of Bioactive Compounds: From Plants to Drug Development*, 2017, pp. 317–331.

[11] N.M. Fanego, D.A. Tacca, and H.E. Olaiz, *Pulse Generator for Electroporation*, 2018, pp. 169–174.

[12] K. Levkov, Y. Linzon, B. Mercadal, A. Ivorra, C.A. González and A. Golberg, High-voltage pulsed electric field laboratory device with asymmetric voltage multiplier for marine macroalgae electroporation, *Innov. Food Sci. Emerg. Technol.* 60 (2020), p. 102288.

[13] S. War Naw, N. Darli Kyaw Zaw, N. Siti Aminah, M. Amin Alamsjah, A. Novi Kristanti, A.S. Nege et al., Bioactivities, heavy metal contents and toxicity effect of macroalgae from two sites in Madura, Indonesia. *J. Saudi Soc. Agric. Sci.* 19 (2020), pp. 528–537.

[14] S. Vidović, J. Vladić, N. Nastić and S. Jokić, Subcritical and supercritical extraction in food by-product and food waste valorization, *Innov. Food Process. Technol.* (2021), pp. 705–721.

[15] E.M. Balboa, A. Moure, and H. Domínguez, Valorization of *Sargassum muticum* biomass according to the biorefinery concept, *Mar. Drugs* 13 (2015), pp. 3745–3760.

[16] E. Conde, A. Moure and H. Domínguez, Supercritical CO2 extraction of fatty acids, phenolics and fucoxanthin from freeze-dried *Sargassum muticum*, *J. Appl. Phycol.* 27 (2015), pp. 957–964.

[17] I. Michalak, B. Górka, P.P. Wieczorek, E. Rój, J. Lipok, B. Łęska, et al., Supercritical fluid extraction of algae enhances levels of biologically active compounds promoting plant growth, *Eur. J. Phycol.* 51 (2016), pp. 243–252.

[18] D.F. Rocher, R.A. Cripwell, and M. Viljoen-Bloom, Engineered yeast for enzymatic hydrolysis of laminarin from brown macroalgae, *Algal Res.* 54 (2021), p. 102233.

[19] H. Li, J. Huang, Y. Xin, B. Zhang, Y. Jin and W. Zhang, Optimization and scale-up of a new photobleaching agar extraction process from *Gracilaria lemaneiformis*, *J. Appl. Phycol.* 21 (2009), pp. 247–254.

3 Bioactive Polysaccharides from Macroalgae

Latifa Tounsi, Faiez Hentati, Olga Babich, Stanislav Sukhikh, Roya Abka Khajouei, Anil Kumar Patel, Reeta Rani Singhania, Imen Fendri, Slim Abdelkafi and Philippe Michaud

CONTENTS

3.1 Introduction 31
3.2 Bioactive Polysaccharides from Red Seaweeds 32
    3.2.1 Extraction and Purification Methods of Red Algae Polysaccharides 32
    3.2.2 Structural and Physicochemical Features of Red Algae Polysaccharides 33
    3.2.3 Biological Activities of Red Algae Polysaccharides 34
        3.2.3.1 Immunomodulatory Activities 34
        3.2.3.2 Antiobesity Activities 35
        3.2.3.3 Antioxidant Activities 35
        3.2.3.4 Anticancer Activities 35
3.3 Bioactive Polysaccharides from Green Seaweeds 36
    3.3.1 Structure of Main Green Algae Polysaccharides 36
        3.3.1.1 Ulvans 36
        3.3.1.2 Arabinans, Galactans and Arabinogalactans 36
        3.3.1.3 Mannans and Xylans 37
    3.3.2 Biological Activity of Green Algae Polysaccharides 37
    3.3.3 The Use of Green Algae Polysaccharides 37
3.4 Bioactive Polysaccharides from Brown Seaweeds 38
    3.4.1 Main Structural Properties of Brown Seaweed Polysaccharides 38
    3.4.2 Bioactivities of Brown Seaweed Polysaccharides 39
3.5 Conclusions and Perspectives 40
References 41

3.1 INTRODUCTION

Polysaccharides are complex macromolecules not directly coded by genome and biosynthesized by all eukaryotic and prokaryotic forms of life, where they are essential for the growth and development of living organisms. Their physiological functions are extremely diverse and depend on their structure and architecture. In vertebrates, polysaccharides participate in a variety of biological processes, including cell-matrix interactions. They form the major structural component of the exoskeletons of marine crustaceans, plants, algae and microorganisms. They play a major role in the complex microbial community structures of flocs and biofilms. They also provide carbon and energy reserves for many types of cells, and they are sometimes excreted. They are recognized for industrial applications and biological properties (Hentati et al., 2020b). Depending on their structure, they can be homopolysaccharides composed of only one monosaccharide species or heteropolysaccharides, sometimes with up to nine to ten different monosaccharides in their structure. Some of them are linear and others ramified, and the kind of glycosidic bonds in their structure strongly impact their solubility in water, even if most of them are hydrophilic molecules due to their hydroxyl group contents. Various substituents such as acyl groups, amino acids or inorganic residues such as sulfates may be attached to these glycans. Seaweeds, also called marine macroalgae, are a heterogeneous group of plants occupying the littoral zone. This term has no taxonomic significance but is currently used to design green algae (Chlorophyta, 7275 species, and Charophyta, 5225 species), brown algae (Ochrophyta, 4486 species) and red seaweeds (Rhodophyta, 7500 species) (www.algaebase.org/). They are multicellular and macrothallic organisms with different shapes, sizes, pigments and contents and occupy different habitats. Some of them are attached to supporting materials such as rocks or to the sea floor through rootlike structures, whereas other float on water surfaces. The taxonomy of these plants is complex and controversial. Ochrophyta belong to kingdom of Chromista, whereas Rhodophyta and Chlorophyta/Charophyta are from kingdom of Plantae. Polysaccharides from seaweeds belong to three classes: food reserve components such as starch and laminaran; cell wall polysaccharides including cellulose and hemicelluloses as found in terrestrial plants; and slimy polysaccharides, sometimes described as pectin-like polysaccharides or the intercellular substance between cells called matrix polysaccharides. These last preserve the hydration of algae, notably in seawater, and allow algae adaptation to water movements. The red seaweeds are composed from the polysaccharidic point of view of starch, sulfated galactans, xylans and cellulose, whereas the brown macroalgae glycans are mainly alginates, fucoidans, laminarans and cellulose. Polysaccharides from green seaweeds are cellulose, mannans, glucomannans, hemicellulose, pectins and sulfated complex heteropolysaccharides. Some macroalgae polysaccharides, also called hydrocolloids or phycocolloids, have long been extracted from wild or cultivated seaweeds for applications as texturing agents, mainly in the food industry. The three main polysaccharides from seaweeds with industrial applications are alginates, agar and carrageenans. More recently, increasing attention has been paid to polysaccharides from macroalgae as a family of biomolecules with various bioactivity with the potential of applications in cosmetic, agricultural and other areas. Indeed, marine organisms are one of the most underutilized biological resources. The extreme diversity of algae that are known to produce large quantities of polysaccharides makes them very attractive for bioprospecting and potential exploitation as commercial sources of bioactive polysaccharides. This chapter aims to view the state of the art of bioactive polysaccharides, including recent developments on their structure elucidation and extraction processes.

3.2 BIOACTIVE POLYSACCHARIDES FROM RED SEAWEEDS

Marine algae like red seaweeds are potential alternative sources of useful bioactive compounds. Besides serving as a food source, red seaweeds (Rhodophyta) are rich in bioactive polysaccharides with diverse physiological and biological activities.

3.2.1 Extraction and Purification Methods of Red Algae Polysaccharides

The characteristics and extraction yields of red sulfated polysaccharides (RSPs) depend on the specific algal species, the growth conditions and the separation methods. The extraction of RSPs is often performed by hydrothermal methods (hot water extraction), which have many drawbacks, such as the use of large volumes of water, long extraction times and low extraction yields. To raise yields and productivity in the food and pharmaceutical industries, innovative and ecological extraction methods, including ultrasound assisted extraction (UAE), ultrasound-microwave assisted extraction (UMAE), supercritical fluid extraction (SFE), pressurized fluid/liquid extraction (PLE), microwave assisted extraction (MAE) and enzyme assisted extraction (EAE), have recently been applied in RSP extraction. For example, the UAE technology enhances the antioxidant capacity of agaran isolated from *Gelidium sesquipedale* compared with hydrothermal extraction, probably due to changes in molecular mass, sulfate content, rheological properties and gel strength (Wu et al., 2022).

During RSP extraction, the co-extracted non-polysaccharide components (e.g. proteins, polyphenols, pigments and other compounds) are usually removed using different organic solvents and enzymatic methods. RSP mixtures can be purified using several methods, including cold ethanol precipitation, membrane separation, biorefinery approach and column chromatography (anion-exchange and gel filtration). To reach large-scale purification of RSPs, the membrane separation method (based on the polysaccharide molecular masses) is expected to support industries in the future thanks to its cost-effectiveness, easy availability of membrane materials, operational flexibility and low energy requirements. High-purity RSPs can be obtained after purification by column chromatography (high cost and difficulties for resin regeneration), which can enable researchers to determine and understand the structure-function relationship of RSPs as well as their bioactivity mechanisms.

3.2.2 Structural and Physicochemical Features of Red Algae Polysaccharides

Red seaweeds are a known industrial source of highly sulfated galactans with texturing and gelling properties such as (i) carrageenans and (ii) agarocolloids. These two biopolymers have similar structures constituted of alternating (1,4)-α-Gal*p* and (1,3)-β-Gal*p* units. The α-Gal*p* units (B unit) are of configuration D in carrageenans and L in agarocolloids, whereas the β-Gal*p* unit (A unit) always belongs to the D series. The (1,4)-α-Gal*p* residues can exist in the 3,6-anhydrogalactose form (3,6-α-AnGal*p*, DA unit) after elimination of the sulfate ester at C-6 by galactose-6-sulfurylases during biosynthesis (Figure 3.1) (Hentati et al., 2020b).

Carrageenans, the third-largest hydrocolloids in the food industry after gelatin and starch, alternately consist of repeating AB units (3-β-D-Gal*p* and 4-α-D-Gal*p*). The B residues can be replaced

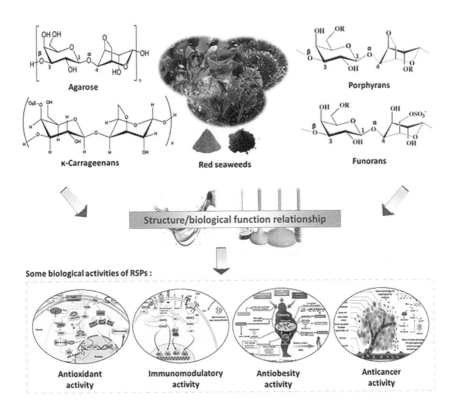

FIGURE 3.1 Structures and some activities of red algae sulfated polysaccharides.

by DA units. These sulfated galactans can be obtained from several species of red algae, such as *Eucheuma, Agardhiella, Furcellaria, Gigartina, Chondrus* and *Hypnea*, and their structural heterogeneity depends on their levels of sulfate and 3,6-α-D-AnGal*p*. There are classified into several forms: κ-(A4S-DA), ι-(A4S-DA2S), λ-(A2S-B2S,6S), ξ-(A2S-B2S), β-(A-DA), α-(A-DA2S), ω-(A6S-DA), μ-(A4S-B6S), ν-(A4S-B2S, 6S), γ-(A-B6S) and δ-(A-B2S,6S), π-(A2S-BP2S), θ-(A2S-DA2S) and ψ-(A6S-B6S) carrageenans (Hentati et al., 2020b). SEC-MALLS chromatography (size exclusion chromatography coupled with refractive index detector, viscometer and multi-angle laser light scattering) is used to estimate the mass-average molecular weight (M_w), number-average molecular mass (M_n), intrinsic viscosity (η) and hydrodynamic (R_h) and gyration (R_g) rays. The molecular configuration of polysaccharides refers to their shape and size in solution, including rigid rods ($R_g/R_h = 2$), random coils ($R_g/R_h = 1.7$) and hard spheres ($R_g/R_h = 0.78$) (Hentati et al., 2019). From Campo et al. (2009), carrageenans presented widely variable M_w values, ranging from 100 to 1400 kDa with flexible coil configuration.

Agarans are sulfated galactans with B residues of the L-series (4-α-L-Gal*p*). There are divided into two groups: agaroids (weak gelling agents) and agars (high gelling agents) (Hentati et al., 2020b), depending on the content of 3–6-α-L-AnGal*p* residues and sulfate groups.

Agars are essentially extracted from *Gelidium, Gracilaria* and *Pterocladia* species, and agarobiose depicts the basic repeating unit of agarose consisting of β-(1,3)-D-Gal*p* and α-(1,4)-L-An-3,6-Gal*p* units. They can be considered mixtures of agarose and agaropectin.

Agarose (M_w ~ 120 kDa and Dp ~ 400) is often heavily substituted in the C-6 position by methyl groups and agaropectin (with the same agarose structure) representing a high branching degree by glycuronate, methyl, pyruvate or sulfate residues.

Agaroids are weakly gelling polymers, which can be divided into (i) funorans and (ii) porphyrans. Funorans are constituted of alternating A (6-*O*-SO$_3^-$) and B (2-*O*-SO$_3^-$) units and are widely used in the adhesives field. Porphyrans (7.86–600 kDa) are highly substituted by sulfate and methyl groups at *O*-4,6 of Gal*p* residues (Bhatia et al., 2015). They are mainly extracted from *Porphyra* species, such as *P. tenera, P. suborbiculata, P. haitanensis* and *P. yezoensis*, and present a flexible chain in aqueous solution (Bhatia et al., 2015).

3.2.3 BIOLOGICAL ACTIVITIES OF RED ALGAE POLYSACCHARIDES

Seaweeds have been used as food raw materials, food ingredients, food supplements and feed. There is now a growing trend in the consumption of red seaweed, thanks to its nutritional properties and health benefits. Several *in vitro* and *in vivo* investigations have demonstrated the effects of RSPs as healthy food supplements, notably in the regulation of disease prevention pathways (Germic et al., 2019).

3.2.3.1 Immunomodulatory Activities

Several *in vitro* and *in vivo* studies have shown that RSPs exhibit immunomodulatory capacities through multiple signaling pathways and targets (Germic et al., 2019; Huang et al., 2019). Oral administration of β/κ-carrageenan (at 100 mg/kg/day) extracted from *Tichocarpus crinitus* to mice could change the motility and morphology of murine peritoneal macrophages as well as increasing serum levels of IFN-γ, IL-1β, IL-4 and IL-12 (Cicinskas et al., 2020). Ren et al. (2017) demonstrated that sulfated polysaccharides from *Gracilariopsis lemaneiformis* could significantly ameliorate the proliferation and pinocytic capability of macrophage RAW264.7 cells and promote the production of ROS, NO, TNF-a and IL-6 by activating mRNA expression of TNF-α, iNOS and IL-6. Immunomodulatory activity was largely correlated with molecular mass and viscosity of RSPs. Enzymatic degraded polysaccharides of *Porphyra haitanensis* (217 kDa) presented more important activities in raising proliferation, phagocytosis and NO secretion by RAW264.7 macrophage cells compared to polysaccharides without hydrolysis (Li et al., 2020).

3.2.3.2 Antiobesity Activities

Obesity increases cardiovascular hazard by raising blood sugar levels and decreasing insulin concentration. Given their positive effects, RSPs could be used to treat glucose metabolism disorders (Wen et al., 2017). Du Preez et al. (2020) showed that mice models fed on high-fat diets (95%) supplemented with 5% κ-carrageenan extracted from *Kappaphycus alvarezii* or with 5% ɩ-carrageenan extracted from *Sarconema filiforme* had decreased body weight, liver fat, abdominal fat, systolic blood pressure and plasma total cholesterol concentrations. Polysaccharides extracted from *G. lemaneiformis* could inhibit cholesterol synthesis by attenuating HMGR and SREBP-2 expression and raising the conversion of cholesterol to bile acids by upregulating the expression of LxRα and CYP7A1 simultaneously (Huang et al., 2019). In obese hamsters, RSPs isolated from *Gelidium amansii* have demonstrated great antiobesity activity due to their ability to regulate plasma adipocytokines by lowering the weight of adipose tissues, body, liver, plasma leptin, triglyceride and total cholesterol levels (Yang et al., 2019). Similarly, sulfated polysaccharides obtained from *Porphyra* spp. could effectively decrease postprandial blood sugar level in rats from 14.70 to 5.35 mmol/L (Zeng et al., 2020).

3.2.3.3 Antioxidant Activities

Oxidative damage (following reactive oxygen species [ROS] production) could lead to the development of diverse pathologies such as neurodegeneration, cardiovascular disease, kidney disease, diabetes and cancer (Hentati et al., 2020b). Numerous studies have reported *in vitro* and *in vivo* antioxidant capacity of polysaccharides from *Gracilaria*, *Porphyra* and others. For example, sulfated polysaccharides from *Gracilaria rubra* have important scavenging activity (~50% at 2.5 mg/mL) against 2,2'-azino-bis(3-ethylbenzothiazoline-6-sulfonic acid (ABTS) and superoxide radicals, and can also inhibit lipid peroxidation (Di et al., 2017). RSPs from *Solieria filiformis* had high scavenging capacities against against 2,2-diphenyl1-picrylhydrazyl (DPPH) and hydroxyl radicals with good dose-dependent reductive capacity (FRAP) (Sousa et al., 2016). Khan et al. (2020) demonstrated that sulfated polysaccharides from *Porphyra haitanensis* (at 2 mg/mL) had ABTS radical scavenging (53.16%), DPPH radical scavenging (34.63%) and hydroxyl radical scavenging (23.80%) potential. *In vivo* investigations have shown that *Kappaphycus alvarezii* carrageenan can ameliorate alloxan-induced oxidative stress response in Wistar albino rats by increasing the activities of antioxidant enzymes, such as SOD, CAT, GST and glutathione in the kidneys of diabetic rats (Sanjivkumar et al., 2020). Similarly, *Gracilaria caudata* polysaccharides presented antioxidant action after intraperitoneal injection of 2,2'-azobis-(2-methylpropionamidine) dihydrochloride (AAPH) in Wistar rats, shown by increased levels of CAT and SOD (Alencar et al., 2019).

3.2.3.4 Anticancer Activities

Anticancer drugs have serious side effects on normal mitotic cells, causing anemia, alopecia, vomiting and nerve changes. To reduce them, investigations have considered RSPs as effective non-toxic antitumor drugs through activation of host immune response. Luo et al. (2015) showed that intratumoral injection of λ-carrageenan inhibits tumor growth of B16-F10 and 4 T1 bearing mice by raising levels of dendric cells, tumor-infiltrating M1 macrophages and activation of CD8$^+$ and CD4$^+$ T lymphocytes in spleen and enhances the secretion of TNF-α and IL-17A. Two different RSPs obtained from *P. yezoensis* and *P. haitanensis* have also been shown to enhance immune response in immunosuppressed mice by regulating the Th1 and Th2 responses to inhibit cancer progression (Fu et al., 2019). Moreover, polysaccharides from *Sarcodia ceylonensis* had *in vivo* anti-tumor activity on Sarcoma-180 bearing mice models by increasing the thymic and splenic indices of the S180 mice to strongly promote the secretion of IL-2, TNF-α and IFN-γ (Fan et al., 2014).

3.3 BIOACTIVE POLYSACCHARIDES FROM GREEN SEAWEEDS

The most common polysaccharides from green macroalgae are starch and cellulose. Therefore, the properties of green algae polysaccharides are more similar to those of land plants than to those of red and brown algae (Mana et al., 2022). Green algae sulfated carbohydrate polymers are bioactive. Moreover, their structural models may vary depending on the type of algae, the place of cultivation and the method of extraction (Hentati et al., 2020b).

3.3.1 Structure of Main Green Algae Polysaccharides

Ulvans, arabinans, galactans, mannans, xylans and their derivatives are the most common bioactive polysaccharides from green seaweeds.

3.3.1.1 Ulvans

The most common water-soluble polysaccharides of green algae, ulvans, have been from the genera *Ulva* and *Enteromorpha* (Priyan Shanura Fernando et al., 2019). These polymers are sulfated heteropolysaccharides composed of repeating units of disaccharides (rhamnose and xylose) and iduronic or glucuronic acids. In this case, the sulfate group, as a rule, is connected to rhamnose (Figure 3.2). The average molecular weight of ulvan ranges from 189 to 8200 kDa (Tanna & Mishra, 2019). Instead of uronic acids, some ulvans may contain xylose or sulfated xylose residues. In this case, disaccharides are called ulvanobioses. Glucuronic or iduronic acids and rhamnose in ulvans occur mainly in the form of aldobiuronic acids. There are two main types of aldobiuronic acids in ulvans. The first type consists of glucuronic acid and sulfated rhamnose, and the other one consists of iduronic acid and sulfated rhamnose, which are linked by (1,4)-glycosidic bonds (Figure 3.2).

Ulvans have a branched complex structure without a definite main chain. The chemical heterogeneity of ulvans leads to an essentially disordered conformation of the biopolymer. Some ulvans can contain higher amounts of rhamnose. The structure of these biopolymers may differ depending on the places and conditions where algae grow. Furthermore, the method of polysaccharide extraction used can have an impact on the structure. Since the aldobiuronic link is resistant to acid hydrolysis and iduronic acid is partially dissolved by acid hydrolysis, precise identification of the monosaccharide composition of ulvans is challenging. However, the structure of ulvans can be characterized using ulvanolytic enzymes that cleave the (1,4) bond between sulfated rhamnose and glucuronic or iduronic acids.

3.3.1.2 Arabinans, Galactans and Arabinogalactans

Green algae are rich in sulfated galactans. This polysaccharide has been isolated, for example, from *Udotea flabellum* (Presa et al., 2018), *Codium isthmocladum* (Sabry et al., 2019) and *Caulerpa*

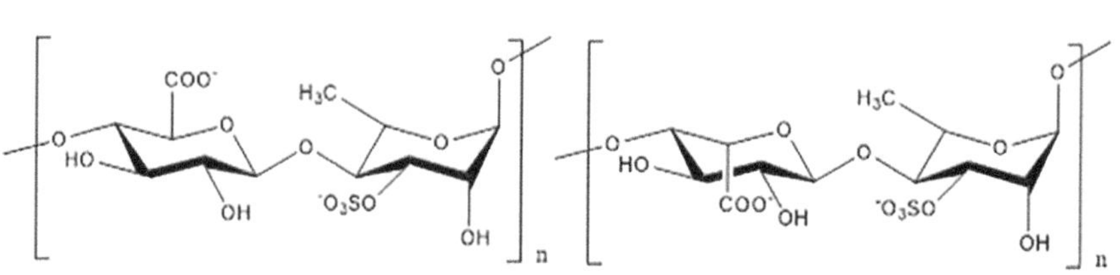

FIGURE 3.2 The structures of the main disaccharide units of ulvans: a) a repeating disaccharide consisting of glucuronic acid and sulfated rhamnose; b) a repeating disaccharide consisting of iduronic acid and sulfated rhamnose.

cupressoides (Barbosa et al., 2020). Sulfated galactans are highly branched sulfated β-D-galactose molecules linked by (1,3) and (1,6) glycosidic bonds. Sulfation mainly occurs at the C-4 and C-6 positions. A distinct feature of green algae galactan is the attachment of pyruvate ketal residues (Priyan Shanura Fernando et al., 2019).

Sulfated arabinans are the less common green algae polysaccharides. These carbohydrates are found in the cell walls of *Codium adhaerens* (Ouahid et al., 2021). These arabinans are highly branched polysaccharides with a backbone of α-(1,5)-arabinan branched by single arabinose units connected by α-(1,2) or α-(1,3) linkages (Ouahid et al., 2021). Algae belonging to orders *Cladophorales* and *Bryopsidales* also contain derivatives of galactans and arabinans, including sulfated xyloarabinogalactans or arabinoxylogalactans. Glucuronoxyloramnogalactans and sulfated rhamnogalactogalacturonans have been isolated from *Ulvales* algae (Hentati et al., 2020a).

3.3.1.3 Mannans and Xylans

Mannans and xylans are polysaccharides found mainly in green algae of the order *Bryopsidales* (Priyan Shanura Fernando et al., 2019). Mannans have also been found in *Codium bursa* (Ouahid et al., 2021). Mannan is classified as a type of hemicellulose which backbones are linear or branched chains composed of D-mannose, D-galactose and D-glucose units. Mannan contains sulfate groups (Priyan Shanura Fernando et al., 2019). Substituted (1,4)-β-D-xylans, similar to those found in terrestrial plants, have been identified mainly in the cell wall matrix of charophytes. (1,3)-β-D-xylans, which are triple helices, were discovered in microfibrils of cell walls of chlorophyte (order Bryopsidales) (Hsieh & Harris, 2019).

3.3.2 BIOLOGICAL ACTIVITY OF GREEN ALGAE POLYSACCHARIDES

Sulfated polysaccharides derived from green algae can exhibit a wide range of biological activities. Their pharmaceutical properties are closely related to the size of the molecules, the ratio of constituent monosaccharides, the content of sulfates and the characteristics of glycosidic bonds. Some green algae polysaccharides stimulate the expression of cytokines in intestinal epithelial cells and have also antimicrobial, immunomodulatory and immunostimulatory activity (Hentati et al., 2020b). In addition, they may exhibit anticancer activity. For example, *Codium intricatum* polysaccharides were able to inhibit the growth of breast cancer cells (MCF-7) (Vasquez & Lirio, 2020).

Ulvans have antioxidant and antihyperlipidemic properties and can act as a natural antioxidant to protect the liver from oxidative stress caused by a high-cholesterol diet (Li et al., 2018). Xyloarabinogalactans or arabinoxylogalactans have significant anticoagulant activity (Hentati, et al., 2020a). The extract of *Codium tomentosum* has a beneficial effect on skin moisture retention due to the presence of polysaccharides in its composition, which have a high water-retaining capacity (Wang et al., 2013). The sulfate content and the molecular weight of the polysaccharides play a major role in water retention capacity.

3.3.3 THE USE OF GREEN ALGAE POLYSACCHARIDES

Algae are being used in a variety of sectors due to their biological activity, including the food, pharmaceutical, and chemical sectors, as well as cosmeceuticals. Polysaccharides of green algae, as well as other types of algae, can be a biologically active component for functional nutrition (Tanna et al., 2019). The special structure of ulvans can be used to produce iduronic acid and sulfated rhamnose, which can be used to synthesize some chemicals, such as flavorants (Priyan Shanura Fernando et al., 2019). In addition, the polysaccharides of Ulva algae are widely used in the production of bioplastics. Ulvans have a film-forming property due to the presence of both hydrophilic (OH, COOH, SO$_4$) and hydrophobic (CH$_3$) functional groups, as well as their polyanionic character. Cellulose can be used as a filler or reinforcing material to form agrocomposites. Ulvans can also be blended with other polymers such as chitosan, polyvinyl chloride, polyvinyl alcohol, poly(3,4-ethylenedioxythiophene) and polyethylene oxide to improve the mechanical properties of the films. Ulvans have gel-forming properties in the presence of divalent calcium, copper and zinc cations, due to which this polysaccharide can be used in drug delivery systems. Ulvans are desirable as raw materials for cosmeceutical products due to their rheological and biofunctional properties (Priyan Shanura Fernando et al., 2019). Mannan, a bioactive and biodegradable polysaccharide found in green algae, also has the properties required for use in the cosmeceutical industry. Glucomannan, a gelling polymer, is used as an emulsifier and thickener in food products (Priyan Shanura Fernando et al., 2019).

3.4 BIOACTIVE POLYSACCHARIDES FROM BROWN SEAWEEDS

Brown algae are composed of a number of biologically active substances such as polysaccharides, carotenoids, proteins and secondary metabolites such as terpenes and polyphenols. Due to the presence of these compounds, they are a well-known source of potent bioactive substances in the medical industry, and these compounds are used for human and animal health applications. They are also approved for the treatment of rheumatic diseases, atherosclerosis, menstrual disorders, hypertension, gastric ulcer, goiter, skin diseases and syphilis (Hentati et al., 2020b). Brown seaweed contains a variety of polysaccharides, including alginate, fucoidans and laminarans, which can be extracted using water or alkaline solutions. The amount of polysaccharides strongly depends on the season, species, age and geographical location. They have prominent biologically active properties such as anti-diabetic, anticancer, antioxidant, anti-viral, antimicrobial, anticoagulant and anti-inflammatory (Hentati et al., 2020b). These biological activities are dominated by the structure of polysaccharides that is specific to each seaweed species.

3.4.1 Main Structural Properties of Brown Seaweed Polysaccharides

Alginates are anionic polysaccharide from brown algae cell walls where they are under salt forms (Na^+, Mg^{2+} and Ca^{2+}). Alginate polymers consists of two monomeric units [β-D-mannuronic acid (M) and α-L-guluronic acid (G)] linked by (1,4) glycosidic bonds in an unbranched chain. The two uronic acids are in pyranosic conformation with $^4C_1$ and $^1C_4$ ring conformations, respectively (Hentati et al., 2019). In the structure of alginate, M and G monomers come together in three forms, homopolymer blocks (MM or GG) and heteropolymer blocks (MGMG). The biological activities of alginates depend on the uronic acid compositions and the relative amount of the three types of blocks. Alginates can form clear gels with divalent cations such as Ca^{+2}, Ba^{+2} and Sr^{+2}, which play an important role in nutrition and medicine (Ghosh et al., 2021).

Laminarans are water-soluble, low molecular weight polysaccharides (5–10 kDa) and the main storage carbon substrates in brown seaweeds. It is composed of a backbone of β-(1,3)-D-glucan ramified by β-(1,6)-D-glucose units (Ermakova et al., 2013). Two types of laminarans are described. The first one has main chains terminated by D-mannitol (M series) residues and the other has chains ending with D-glucose residues (G series). Environmental factors such as nutritive salts, age and species of seaweed affect the ratio of the two types of laminarans as well as their structure.

Fucoidans are sulfated and water-soluble polysaccharides from the cell wall of brown seaweeds containing a significant portion of L-fucose (Sanjeewa et al., 2017). The structure of fucoidans are complex in nature, and due to differences in structure (fucose bond, sulfate position and sugar composition), their diversity is very high. However, most algal fucoidans contain a random (1,3)/(1,4) linear backbone of L-fucopyranose residues with less than 10% of other monosaccharides (mannose, galactose, glucose, arabinose, uronic acids and xylose). Sulfate ester groups are arbitrarily substituted at position 2, 3 or 4 of fucopyranoses, making it a highly heterogeneous polymer (Sanjeewa et al., 2017, 2019)

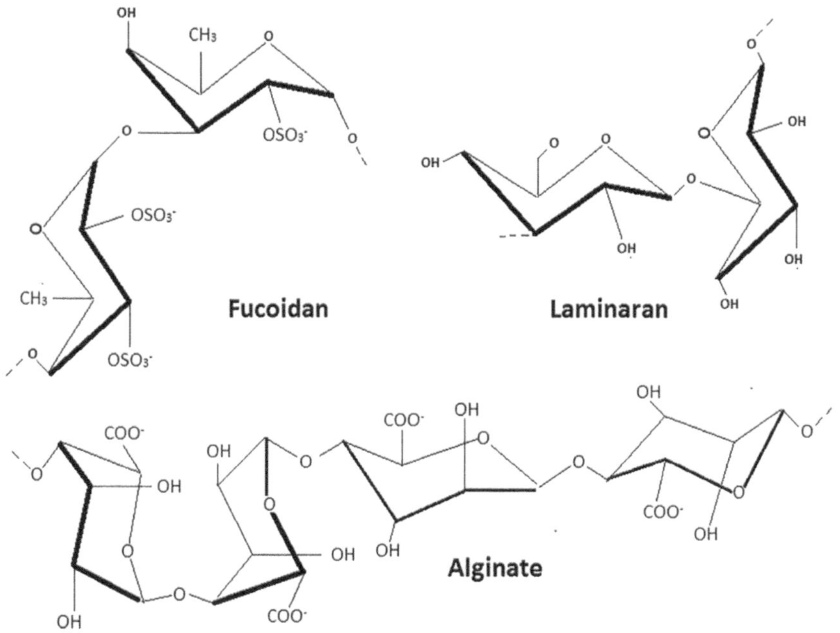

FIGURE 3.3 Structure of fucoidan, laminaran and alginate.

3.4.2 BIOACTIVITIES OF BROWN SEAWEED POLYSACCHARIDES

Alginate dressings are used for wound care due to their excellent absorbency, antimicrobial properties and ability to create a moist environment by soaking secretions as well as forming non-stick gels. Alginate composites have antioxidant and bactericidal activities and enhance tissue repair through targeted delivery of cells and proteins in the body. The rate of antioxidant activity depends on the availability of a number of the carboxylic groups (-COOH) of the uronic acid unit (Hentati et al., 2019, 2020b), so molecular weight and M/G ratio are important factors in controlling the antioxidant properties of sodium alginate. In low molecular weight alginates, antioxidant properties due to increased chain flexibility or reduced compact structure as well as formation more hydroxyl groups (due to broken glycosidic bonds in skeletal structure) are higher. Diaxial bonding in G blocks leads to impaired rotation around the glycosidic bond and increases the flexibility of G blocks, thus affecting the availability of sodium alginate hydroxyl groups. Putting alginate in the diet may reduce the risk of type 2 diabetes (especially in high-risk populations), obesity and possibly cardiovascular disease. It also reduces systemic risk factors in these patients. This effect of alginate has been attributed to a reduction in intestinal absorption of plasma glucose/cholesterol and prolonging gastric emptying (leads to increased satiety) (Brownlee et al., 2005). Alginates protect the stomach from hydrochloride-ethanol–induced ulcers in rats, possibly due to the gel-forming capacity on the surface of the gastric mucosa by the existence of monocarbonic acid (Ammar et al., 2018).

Laminarans have interesting bioactive properties such as immunoregulatory, antiapoptotic, antioxidant, anti-tumor, anti-inflammatory, anti-coagulation, strong immunomodulatory, radio-protective and anticancer (Sanjeewa et al., 2017). The sulfated derivatives of laminarans compared to native laminarans are more effective in inhibition of proliferation, colony formation and migration and induce apoptosis of cancer cells (Malyarenko et al., 2017). Malyarenko et al. (2017) have shown that sulfated laminarans from *F. evanescens* have high anticancer activity *in vitro* and inhibit the activity of the matrix metalloproteinase-2 and -9 (MMP-2 and MMP-9). As a result, they prevent the migration of breast cancer cells. These properties are probably related to the high degree of sulfation, the content of glycosidic bonds and the molecular structure characteristics of these polysaccharides, so they may be one of the therapeutic approaches to treat cancer (Malyarenko et al., 2017). Aqueous free radicals are produced by radiation on water of body tissues, which damage DNA and RNA by reacting with them. Laminarans have antioxidant activity, so they can minimize or eliminate free radical damage and prevent X-ray toxicity (Malyarenko et al., 2019).

Fucoidans are highly bioactive seaweed with many promising physiological activities that have attracted the attention of many industries around the world (Sanjeewa et al., 2019). They have been widely used because of their beneficial and interesting biological activities such as anti-inflammatory, immunomodulatory, antioxidant, anticoagulant, antiviral, anticancer, skin protecting and antiaging activities and protection of the stomach (Sanjeewa et al., 2019). The strong biological properties of fucoidans are associated with their high degree of sulfation, fine structure and molecular weight (Wijesinghe & Jeon, 2011). Fucoidans stimulate the immune system in a variety of ways, such as changing cell surface properties or directly inhibiting oral intake, modulating the immune system by helping recovery of cells or acting as anti-inflammatory agents. Fucoidans stimulate macrophages and splenocytes to enhance the release of cytokines and chemokines (Ghosh et al., 2021). They have antiviral properties by binding to viral coat glycoproteins and preventing the virus from binding to the host cell (Nishino et al., 1991). In some types of cancer, fucoidans can undergo metastatic processes including epithelial–mesenchymal transition (EMT), migration, prevention of invasive processes and mesenchymal–epithelial transition (MET) (Reyes et al., 2020). In addition, they can suppress the development of tumor cells by increasing the immune system (Wijesinghe & Jeon, 2011). The molecular properties of fucoidans and the possibility of chemical or enzymatic changes in their structure lead to their therapeutic use (especially as an aid to increase the effectiveness of chemotherapy). The molecular versatility of fucoidans and the possibility of preparing nanoparticles with their help have shown their potential to improve drug delivery efficiency to tumors or to achieve a synergistic effect with other drugs (Reyes et al., 2020). Fucoidans strongly prevent coagulation activity by increasing antithrombin III–mediated coagulation factor inhibition (Wijesinghe & Jeon, 2011).

3.5 CONCLUSIONS AND PERSPECTIVES

In the context of blue biotechnologies, seaweeds are today a source of high-value molecules, and more and more pharmaceuticals are isolated from them. They have been largely exploited for more than a century for their hydrocolloid content. These last found numerous applications in food for their texturing properties, leading to the cultivation of industrially important species. Matrix water-soluble polysaccharides from seaweeds have different structures not found in terrestrial plants. These highly specific structures are correlated with original rheological and physical properties. Recently these marine hydrocolloids, also called phycocolloids, have been recognized for many biological applications in development in various industrial fields. The main drawbacks limiting their development as therapeutic and/or bioactive agents despite their identified biological properties, sometimes higher than existing drugs, is their structural heterogeneity. Polysaccharides from seaweeds as those from other organisms are always mixture of macromolecules having a generic structure not fully representative of each biopolymer present in the polysaccharidic extract. Indeed, the molecular weight, repetition of monosaccharides and location of non-sugar groups vary within the same batch. This heterogeneity is difficult to explain to regulatory agencies, which want a pure molecule with a defined structure for all drugs. To better explain this point, the example of anticoagulant drugs is probably the best. Heparin extracted from animals has been used for several decades even though some sulfated polysaccharides and notably sulfated galactans from algae have equivalent or higher activity. They are not able to compete with this polysaccharide, as they are a mixture of sulfated galactans with medium molecular weight, sulfation degree and anhydrogalactose contents. Note that if heparin had been introduced to the anticoagulant market today, it would have faced the same problem. The future for bioactive polysaccharides will probably come through the development of extraction, purification and analytical methods to allow preparation of better-characterized mixtures of polysaccharides. Their use as bioactive hydrogels or bioactive particles after 3D printing or linkage with non-polysaccharide nanoparticles will also offer new fields of application for them.

REFERENCES

Alencar, P. O. C., Lima, G. C., Barros, F. C. N., Costa, L. E. C., Ribeiro, C. V. P. E., Sousa, W. M., Sombra, V. G., Abreu, C. M. W. S., Abreu, E. S., Pontes, E. O. B., Oliveira, A. C., de Paula, R. C. M., & Freitas, A. L. P. (2019). A novel antioxidant sulfated polysaccharide from the algae *Gracilaria caudata*: In vitro and in vivo activities. *Food Hydrocolloids*, *90*, 28–34. https://doi.org/10.1016/j.foodhyd.2018.12.007

Ammar, H. H., Lajili, S., Sakly, N., Cherif, D., Rihouey, C., Le Cerf, D., Bouraoui, A., & Majdoub, H. (2018). Influence of the uronic acid composition on the gastroprotective activity of alginates from three different genus of Tunisian brown algae. *Food Chemistry*, *239*, 165–171. https://doi.org/10.1016/j.foodchem.2017.06.108

Barbosa, J. D. S., Sabry, D. A., Silva, C. H. F., Gomes, D. L., Santana-Filho, A. P., Sassaki, G. L., & Rocha, H. A. O. (2020). Immunostimulatory effect of sulfated galactans from the green seaweed *Caulerpa cupressoides* var. flabellata. *Marine Drugs*, *18*(5), 234. https://doi.org/10.3390/md18050234

Bhatia, S., Sharma, K., Nagpal, K., & Bera, T. (2015). Investigation of the factors influencing the molecular weight of porphyran and its associated antifungal activity. *Bioactive Carbohydrates and Dietary Fibre*, *5*(2), 153–168. https://doi.org/10.1016/j.bcdf.2015.03.005

Brownlee, I. A., Allen, A., Pearson, J. P., Dettmar, P. W., Havler, M. E., Atherton, M. R., & Onsøyen, E. (2005). Alginate as a source of dietary fiber. *Critical Reviews in Food Science and Nutrition*, *45*(6), 497–510. https://doi.org/10.1080/10408390500285673

Campo, V. L., Kawano, D. F., Silva, D. B. da, & Carvalho, I. (2009). Carrageenans: Biological properties, chemical modifications and structural analysis—a review. *Carbohydrate Polymers*, *77*(2), 167–180. https://doi.org/10.1016/j.carbpol.2009.01.020

Cicinskas, E., Kalitnik, A. A., Karetin, Y. A., Mohan Ram, M. S. G., Achary, A., & Kravchenko, A. O. (2020). Immunomodulating properties of carrageenan from *Tichocarpus crinitus*. *Inflammation*, *43*(4), 1387–1396. https://doi.org/10.1007/s10753-020-01216-x

Di, T., Chen, G., Sun, Y., Ou, S., Zeng, X., & Ye, H. (2017). Antioxidant and immunostimulating activities in vitro of sulfated polysaccharides isolated from *Gracilaria rubra*. *Journal of Functional Foods*, *28*, 64–75. https://doi.org/10.1016/j.jff.2016.11.005

du Preez, R., Paul, N., Mouatt, P., Majzoub, M. E., Thomas, T., Panchal, S. K., & Brown, L. (2020). Carrageenans from the red seaweed *Sarconema filiforme* attenuate symptoms of diet-induced metabolic syndrome in rats. *Marine Drugs*, *18*(2), 97. https://doi.org/10.3390/md18020097

Ermakova, S., Men'shova, R., Vishchuk, O., Kim, S.-M., Um, B.-H., Isakov, V., & Zvyagintseva, T. (2013). Water-soluble polysaccharides from the brown alga *Eisenia bicyclis*: Structural characteristics and antitumor activity. *Algal Research*, *2*(1), 51–58. https://doi.org/10.1016/j.algal.2012.10.002

Fan, Y., Lin, M., Luo, A., Chun, Z., & Luo, A. (2014). Characterization and antitumor activity of a polysaccharide from *Sarcodia ceylonensis*. *Molecules*, *19*(8), 10863–10876. https://doi.org/10.3390/molecules190810863

Fu, L., Qian, Y., Wang, C., Xie, M., Huang, J., & Wang, Y. (2019). Two polysaccharides from *Porphyra* modulate immune homeostasis by NF-κB-dependent immunocyte differentiation. *Food & Function*, *10*(4), 2083–2093. https://doi.org/10.1039/C9FO00023B

Germic, N., Frangez, Z., Yousefi, S., & Simon, H.-U. (2019). Regulation of the innate immune system by autophagy: Monocytes, macrophages, dendritic cells and antigen presentation. *Cell Death & Differentiation*, *26*(4), 715–727. https://doi.org/10.1038/s41418-019-0297-6

Ghosh, T., Singh, R., Nesamma, A. A., & Jutur, P. P. (2021). Marine polysaccharides: Properties and applications. In Inamuddin, M. I. Ahamed, R. Boddula, & T. Altalhi (Eds.), *Polysaccharides* (1st ed., pp. 37–60). Wiley. https://doi.org/10.1002/9781119711414.ch3

Hentati, F., Delattre, C., Gardarin, C., Desbrières, J., Le Cerf, D., Rihouey, C., Michaud, P., Abdelkafi, S., & Pierre, G. (2020a). Structural features and rheological properties of a sulfated xylogalactan-rich fraction isolated from Tunisian red seaweed *Jania adhaerens*. *Applied Sciences*, *10*(5), 1655. https://doi.org/10.3390/app10051655

Hentati, F., Tounsi, L., Djomdi, D., Pierre, G., Delattre, C., Ursu, A. V., Fendri, I., Abdelkafi, S., & Michaud, P. (2020b). Bioactive polysaccharides from seaweeds. *Molecules*, *25*(14), 3152. https://doi.org/10.3390/molecules25143152

Hentati, F., Ursu, A. V., Pierre, G., Delattre, C., Trica, B., Abdelkafi, S., Djelveh, G., Dobre, T., & Michaud, P. (2019). Production, extraction and characterization of alginates from seaweeds. In *Handbook of Algal Technologies and Phytochemicals*. CRC Press.

Hsieh, Y. S. Y., & Harris, P. J. (2019). Xylans of red and green algae: What is known about their structures and how they are synthesised? *Polymers*, *11*(2), 354. https://doi.org/10.3390/polym11020354

Huang, L., Shen, M., Morris, G. A., & Xie, J. (2019). Sulfated polysaccharides: Immunomodulation and signaling mechanisms. *Trends in Food Science & Technology*, *92*, 1–11. https://doi.org/10.1016/j.tifs.2019.08.008

Khan, B. M., Qiu, H.-M., Xu, S.-Y., Liu, Y., & Cheong, K.-L. (2020). Physicochemical characterization and antioxidant activity of sulphated polysaccharides derived from *Porphyra haitanensis*. *International Journal of Biological Macromolecules*, *145*, 1155–1161. https://doi.org/10.1016/j.ijbiomac.2019.10.040

Li, W., Jiang, N., Li, B., Wan, M., Chang, X., Liu, H., Zhang, L., Yin, S., Qi, H., & Liu, S. (2018). Antioxidant activity of purified ulvan in hyperlipidemic mice. *International Journal of Biological Macromolecules*, *113*, 971–975. https://doi.org/10.1016/j.ijbiomac.2018.02.104

Li, Y., Huo, Y., Wang, F., Wang, C., Zhu, Q., Wang, Y., Fu, L., & Zhou, T. (2020). Improved antioxidant and immunomodulatory activities of enzymatically degraded *Porphyra haitanensis* polysaccharides. *Journal of Food Biochemistry*, *44*(5). https://doi.org/10.1111/jfbc.13189

Luo, M., Shao, B., Nie, W., Wei, X.-W., Li, Y.-L., Wang, B.-L., He, Z.-Y., Liang, X., Ye, T.-H., & Wei, Y.-Q. (2015). Antitumor and adjuvant activity of λ-carrageenan by stimulating immune response in cancer immunotherapy. *Scientific Reports*, *5*(1), 11062. https://doi.org/10.1038/srep11062

Malyarenko, O. S., Usoltseva, R. V., Shevchenko, N. M., Isakov, V. V., Zvyagintseva, T. N., & Ermakova, S. P. (2017). In vitro anticancer activity of the laminarans from Far Eastern brown seaweeds and their sulfated derivatives. *Journal of Applied Phycology*, *29*(1), 543–553. https://doi.org/10.1007/s10811-016-0915-3

Malyarenko, O. S., Usoltseva, R. V., Zvyagintseva, T. N., & Ermakova, S. P. (2019). Laminaran from brown alga *Dictyota dichotoma* and its sulfated derivative as radioprotectors and radiosensitizers in melanoma therapy. *Carbohydrate Polymers*, *206*, 539–547. https://doi.org/10.1016/j.carbpol.2018.11.008

Nishino, T., Kiyohara, H., Yamada, H., & Nagumo, T. (1991). An anticoagulant fucoidan from the brown seaweed *Ecklonia kurome*. *Phytochemistry*, *30*(2), 535–539. https://doi.org/10.1016/0031-9422(91)83722-W

Ouahid, E. A., Mohamed, R., & Soufiane, F. (2021). Green seaweed polysaccharides: Inventory of Nador Lagoon in North East Morocco. In Inamuddin, M. I. Ahamed, R. Boddula, & T. Altalhi (Eds.), *Polysaccharides* (1st ed., pp. 163–175). Wiley. https://doi.org/10.1002/9781119711414.ch8

Presa, F. B., Marques, M. L. M., Viana, R. L. S., Nobre, L. T. D. B., Costa, L. S., & Rocha, H. A. O. (2018). The protective role of sulfated polysaccharides from green seaweed *Udotea flabellum* in cells exposed to oxidative damage. *Marine Drugs*, *16*(4), 135. https://doi.org/10.3390/md16040135

Priyan Shanura Fernando, I., Kim, K.-N., Kim, D., & Jeon, Y.-J. (2019). Algal polysaccharides: Potential bioactive substances for cosmeceutical applications. *Critical Reviews in Biotechnology*, *39*(1), 99–113. https://doi.org/10.1080/07388551.2018.1503995

Ren, Y., Zheng, G., You, L., Wen, L., Li, C., Fu, X., & Zhou, L. (2017). Structural characterization and macrophage immunomodulatory activity of a polysaccharide isolated from *Gracilaria lemaneiformis*. *Journal of Functional Foods*, *33*, 286–296. https://doi.org/10.1016/j.jff.2017.03.062

Reyes, M. E., Riquelme, I., Salvo, T., Zanella, L., Letelier, P., & Brebi, P. (2020). Brown seaweed fucoidan in cancer: Implications in metastasis and drug resistance. *Marine Drugs*, *18*(5), 232.

Sabry, D. A., Cordeiro, S. L., Ferreira Silva, C. H., Cunha Farias, E. H., Sassaki, G. L., Nader, H. B., & Oliveira Rocha, H. A. (2019). Pharmacological prospection and structural characterization of two purified sulfated and pyruvylated homogalactans from green algae *Codium isthmocladum*. *Carbohydrate Polymers*, *222*, 115010. https://doi.org/10.1016/j.carbpol.2019.115010

Sanjeewa, K. K. A., Jayawardena, T. U., Kim, S.-Y., Kim, H.-S., Ahn, G., Kim, J., & Jeon, Y.-J. (2019). Fucoidan isolated from invasive *Sargassum horneri* inhibit LPS-induced inflammation via blocking NF-κB and MAPK pathways. *Algal Research*, *41*, 101561. https://doi.org/10.1016/j.algal.2019.101561

Sanjeewa, K. K. A., Lee, J.-S., Kim, W.-S., & Jeon, Y.-J. (2017). The potential of brown-algae polysaccharides for the development of anticancer agents: An update on anticancer effects reported for fucoidan and laminaran. *Carbohydrate Polymers*, *177*, 451–459. https://doi.org/10.1016/j.carbpol.2017.09.005

Sanjivkumar, M., Chandran, M. N., Suganya, A. M., & Immanuel, G. (2020). Investigation on bio-properties and in-vivo antioxidant potential of carrageenans against alloxan induced oxidative stress in Wistar albino rats. *International Journal of Biological Macromolecules*, *151*, 650–662. https://doi.org/10.1016/j.ijbiomac.2020.02.227

Sousa, W. M., Silva, R. O., Bezerra, F. F., Bingana, R. D., Barros, F. C. N., Costa, L. E. C., Sombra, V. G., Soares, P. M. G., Feitosa, J. P. A., de Paula, R. C. M., Souza, M. H. L. P., Barbosa, A. L. R., & Freitas, A. L. P. (2016). Sulfated polysaccharide fraction from marine algae *Solieria filiformis*: Structural characterization, gastroprotective and antioxidant effects. *Carbohydrate Polymers*, *152*, 140–148. https://doi.org/10.1016/j.carbpol.2016.06.111

Tanna, B., & Mishra, A. (2019). Nutraceutical potential of seaweed polysaccharides: Structure, bioactivity, safety, and toxicity. *Comprehensive Reviews in Food Science and Food Safety*, *18*(3), 817–831. https://doi.org/10.1111/1541-4337.12441

Vasquez, R. D., & Lirio, S. (2020). Content analysis, cytotoxic, and anti-metastasis potential of bioactive polysaccharides from green alga *Codium intricatum* okamura. *Current Bioactive Compounds*, *16*(3), 320–328. https://doi.org/10.2174/1573407214666181019124339

Wang, J., Jin, W., Hou, Y., Niu, X., Zhang, H., & Zhang, Q. (2013). Chemical composition and moisture-absorption/retention ability of polysaccharides extracted from five algae. *International Journal of Biological Macromolecules*, *57*, 26–29.

Wen, L., Zhang, Y., Sun-Waterhouse, D., You, L., & Fu, X. (2017). Advantages of the polysaccharides from *Gracilaria lemaneiformis* over metformin in antidiabetic effects on streptozotocin-induced diabetic mice. *RSC Advances*, *7*(15), 9141–9151. https://doi.org/10.1039/C6RA26970B

Wijesinghe, W. A. J. P., & Jeon, Y.-J. (2011). Biological activities and potential cosmeceutical applications of bioactive components from brown seaweeds: A review. *Phytochemistry Reviews*, *10*(3), 431–443. https://doi.org/10.1007/s11101-011-9214-4

Wu, D.-T., He, Y., Fu, M.-X., Gan, R.-Y., Hu, Y.-C., Peng, L.-X., Zhao, G., & Zou, L. (2022). Structural characteristics and biological activities of a pectic-polysaccharide from okra affected by ultrasound assisted metal-free Fenton reaction. *Food Hydrocolloids*, *122*, 107085. https://doi.org/10.1016/j.foodhyd.2021.107085

Yang, T.-H., Chiu, C.-Y., Lu, T.-J., Liu, S.-H., & Chiang, M.-T. (2019). The anti-obesity effect of polysaccharide-rich red algae (*Gelidium amansii*) hot-water extracts in high-fat diet-induced obese hamsters. *Marine Drugs*, *17*(9), 532. https://doi.org/10.3390/md17090532

Zeng, A., Yang, R., Yu, S., & Zhao, W. (2020). A novel hypoglycemic agent: Polysaccharides from laver (*Porphyra* spp.). *Food & Function*, *11*(10), 9048–9056. https://doi.org/10.1039/D0FO01195A

4 Marine Microalgae in Food and Health Applications

Tirath Raj, K Chandrasekhar, Raj Morya and Sang-Hyoun Kim

CONTENTS

4.1	Introduction	45
4.2	Marine Microalgal Groups for Food and Health Applications	46
	4.2.1 Photosynthetic Bacteria	50
	4.2.2 Eukaryotic Microalgae	50
4.3	Biochemical Composition of Marine Microalgae	51
	4.3.1 Proteins	52
	4.3.2 Carbohydrates	52
	4.3.3 Lipids	53
	4.3.4 Vitamins	53
	4.3.5 Polyphenols	54
	4.3.6 Phytosterols	54
	4.3.7 Carotenoids	54
4.4	Microalgae as Food Supplement	55
4.5	Biological Activity from Microalgae	55
	4.5.1 Anticancer Activity	55
	4.5.2 Anti-Inflammatory Activity	55
	4.5.3 Anti-Angiogenic Activity	57
	4.5.4 Antioxidant Activity	57
	4.5.5 Neuroprotective Activity	57
4.6	Summary and Conclusions	58
References		58

4.1 INTRODUCTION

Algae are biochemically and physiologically quite similar to the rest of the plant kingdom: they contain chlorophyll, synthesize identical proteins and carbohydrates, and have similar metabolic pathways. Several algae, including *Euglenophytes*, *Dinophytes*, and *Ochrophytes*, have lost their photosynthetic ability and now exist as saprophytes or parasites. Green algae, for example, has more than a hundred heterotrophic species to its name, and there are other groups represented as well. The absence of an embryo and a multicellular envelope surrounding the sporangia and gametangia of algae separates them from other photosynthetic plants (except for freshwater green algae, *Charophytes*) (Singh et al., 2020). The algae provide oxygen and reduce chemicals such as maltose, glucose, amino acids, or glycerol to the partner, which provides inorganic materials and a stable habitat to the algae (Enamala et al., 2021). Marine *Chlorophyceae*, for example, may be found within *Platyhelminthes* and nudibranchs. Some dinoflagellates, known as *zooxanthellae*, coexist alongside sponges, jellyfish, and marine coelenterates. *Symbiodinium*, a dinoflagellate, is vital in the creation of coral reefs because it drives colony calcification.

On the Pacific coasts of Asia and South America, the practical utilization of algae as food, animal feed, or fertilizer has a long history. Other places, such as Europe, are seeing a significant increase

DOI: 10.1201/9781003326946-4

in its usage. Some calcified red algae species produce large fields on the seafloor (Kholssi et al., 2021). They're gathered, crushed, and used to reduce the acidity of agricultural soils. Recently, vast probability of the marine algal biomasses for selective uses in fields as varied as human food and aquaculture feed, fertilizers and waste treatment, anti-inflammatory and analgesic compounds, were also investigated (Hussain et al., 2021). In fact, a few unicellular algae, including *Porphyridium* and *Rhodella* (rhodophytes), and cyanobacteria, including *Arthrospira*, can generate sulphated polysaccharides and have been widely accepted for *in vivo* and *in vitro* antiviral components, nutraceutical applications and tumor growth inhibitors.

Chlorophyll-a, -d, and -f pigments, in addition to phycobiliproteins, phycocyanin, and phycoerythrin, are all found in cyanobacteria (Russell et al., 2022). Glaucophytes have chlorophyll-a and use phycobiliproteins to gather light. Chlorophyta includes chlorophyll-a and b, along with various colored pigmented carotenoids, such as carotene and other xanthophylls (Kholssi et al., 2021). Rhodophyte algae's principal pigments mainly include phycoerythrin and phycocyanin elements, which hide chlorophyll-a. Although red algae synthesize a variety of carotenes varying chain length and xanthophyll light-harvesting pigments except chlorophylls. While consuming microalgae in their entirety, one gains access to a wide range of chemicals synthesized through unicellular marine algae that have demonstrated a range of human health benefits with improvement to overall health (Russell et al., 2022). These compounds include polyunsaturated fatty acids (PUFAs), sterols, pigments, amino acids, enzyme secretion, vitamins (mostly A, B, C), and numerous others. Cyanobacteria and eukaryotic microalgae additionally generate a large number of secondary metabolites, most of which are potentially beneficial, including antioxidants, and also a large number of chemicals that are generally classified as poisons (Barkia et al., 2019). These bioactive molecules are synthesized over the shikimic acid, mevalonic/non-mevalonic acid, and acetate pathways and include, but are not limited to, anatoxins, brevetoxins, microcystins, and prymnesins. In excess of 200 bioactive metabolites have been predicted for cyanobacteria and many more for eukaryotic microalgae. A lack of assays and known detection methods for microalgal metabolites suggests that more research is needed in this area.

Thus, microalgae are being hailed as the next generation of robust microorganism to boot biotechnological developments with recent technological, genetic engineering, and microbiological tools for production of renewable carbon fuels, chemicals, treatment of industrial wastewater (Enamala et al., 2021; Yadavalli et al., 2020), and animal and human food supplements. According to the most recent economic feasibility studies, limited biomass productivity and excessive pricing of commercial cultivation, harvesting, and manufacturing biofuels is not economically viable unless there is development of integrated microalgal production with current biorefinery platforms for commercial production of high added value chemical and material production. The utilization of microalgal biomass in the production of health-promoting products like nutraceuticals and functional meals is a fast-increasing industry (Table 4.1). For instance, the global carotenoid demand was predicted to increase from $1.24 billion in 2016 to $1.53 billion in 2021 (Borowitzka, 2013). Hence, microalgae can be classified as a high nutrition supplement for human food, preparation of bioactive medicinal products, and nutraceutical purposes.

4.2 MARINE MICROALGAL GROUPS FOR FOOD AND HEALTH APPLICATIONS

Most of the research on microalgal species has focused on cyanobacteria, spirulina, *Chlorella*, and *Dunaliella* spp., which are all being exploited as nutraceuticals and health food additives, and *Haematococcus pluvialis* and other cyanobacterial sources of carotene and astaxanthin (Eltanahy and Torky, 2021). For example, spirulina is a prokaryotic cyanobacterium, a filamentous microalga that has been further studied and cultivated for large-scale production of vitamins, nutraceuticals, dye, and food supplements. Spirulina is very rich in protein (60–70%), beta-carotene, vitamins, phycocyanin, chlorophyll, omega fatty acids, and minerals. Microalgal application in aquaculture feed additives for a fish fry is another reputable market for algal biomass; feed additives supply the critical long-chain omega 3-fatty acids (LC-PUFAs) and vitamins present in marine microalgae.

TABLE 4.1
Microalgal Products and Manufacturers for Food and Nutrition Applications

Microalgae (form)	Major Producer and Country	Company Details	Total Production per Year (tons/year)	Website
Spirulina Arthrospira (Whole, dried microalgae)	Earthrise Nutraceuticals, USA	Earthrise spirulina was the first company to produce spirulina in the United States, California. An agreement was struck with Dainippon Ink and Chemicals (DIC) in 1981, a Japanese business that had just started cultivating spirulina in Thailand, to help Earthrise develop micro-algae for food, biochemicals, and medicines. Dainippon Ink and Chemicals purchased Earthrise Nutritionals in 2005.	450	www.earthrise.com
	Cyanotech Corporation, USA	Cyanotech, a 96-acre facility on Hawaii's Kailua-Kona Coast, where these items are made, has various advantages. For the company's Ocean-Chill Drying technology and for microalgae cultures that rely on trace nutrients, this site is ideal since it provides access to cold deep ocean water taken from an offshore depth of 2,000 feet.	360	www.cyanotech.com
	Hainan DIC Microalgae, China	DIC has been known to produce organic pigments and synthetic resin operations while fostering world-class associated core technologies.	350	www.dic-global.com
	Japan Algae Co., Ltd., Japan	Japan Algae Co., Ltd., was created on the basis of biotechnology and has grown steadily as a venture firm based on two fields: organic germanium, with a manufacturing patent, and spirulina.	30 to 100	www.sp100.com
	Parry Nutraceuticals, India	Parry Nutraceuticals is a subsidiary of E.I.D. Parry (I) Ltd. and part of the USD 3.8 billion Murugappa group. It joined the Murugappa Group in 1981, and its activities currently include sugar and microalgal health supplements.	Greater than 170	www.parrynutraceuticals.com

(*Continued*)

TABLE 4.1 (CONTINUED)

Microalgae (form)	Major Producer and Country	Company Details	Total Production per Year (tons/year)	Website
	FEMICO, Taiwan	Far East Microalgae Industries, Co., Ltd. (FEMICO), was established in 1976. More than 100,000 tons of chlorella and spirulina are producing vitamin-rich microalgae (chlorella, spirulina, and red algae) powder, SPA, and cosmetics items.	50 to 150	www.femico.com.tw
	Nan Pao International, Taiwan	Tang Qing Yun, Taiwan, launched Nan Pao in 1963 for production of adhesives, footwear adhesives, powder coatings, hot-melt adhesives, and building components. Nan Pao presently has commercial developments and subsidiaries in Taiwan, China, Vietnam, Indonesia, Thailand, Malaysia, and India, with over 2,800 workers globally.	70	www.nanpao.com
	Biorigin, Switzerland	Biorigin is a Brazilian corporation company that provides natural components for taste improvement, salt reduction, food preservatives, nutrient enrichment, and health improvements using biotechnologies. Biorigin is a division of Zilor, an ethanol, sugar, and energy producer with more than 70 years of expertise in Brazil. Since its inception, the company has expanded and strengthened its worldwide footprint by acquiring PTX Food Corp. in the United States and Immunocorp Animal Health in Norway. Biorigin relies on cutting-edge technology and creative thinking to bring new products to market. Biorigin is unique in that it maintains complete control over the whole manufacturing process.	60	www.biorigin.net

TABLE 4.1 (CONTINUED)

Microalgae (form)	Major Producer and Country	Company Details	Total Production per Year (tons/year)	Website
	TAAU Australia Pty Ltd., Australia	In late 1996, the Northern Territory of Australia and TAAU Australia Pvt. Ltd commercially started cultivating spirulina in Darwin for human consumption algae research.	50 to 60	www.australianspirulina.com.au
Chlorella (Whole, dried microalgae)	Roquette Klötze, Germany	The major B2B platform in the DACH area, wlw ("*Wer liefert was*"), assists small and medium-sized businesses in Germany, Austria, and Switzerland in using digital channels for sales and marketing and thereby expanding their Internet reach.	130 to 150	www.wlw.de
	Earthrise Nutritionals, USA	Earthrise Nutritional started producing blue-green algal spirulina in 1976 in the California desert and subsequently changed its name to reflect this.	Data not available	www.earthrise.com
β-carotene from *Dunaliella*	Cognis Nutrition & Health Co., Australia	William Reed's Nutra Ingredients brand has been the premier online news source for the nutrition industry from last 19 years.	Data not available	www.nutraingredients.com
	Nature Beta Technology Ltd., Israel	Nature Beta Technologies (NBT) is a subsidiary of Nikken Sohonsha Corporation in Japan and cultivates *Dunaliella bardawil* microalgae, a source of beta carotene containing 50% of the 9-cis isomer.	Data not available	www.dnb.com
Astaxanthin from *Haematococcus*	Cyanotech Corp., USA	Founded in 1983, Cyanotech Corporation produced high-quality microalgae products for health and nutrition while operating in a sustainable, dependable, and environmentally conscious manner.	Data not available	www.cyanotech.com
	Fuji Chemical Industries, Japan	Fuji Chemical Industries was established in 1946 with roots in the Confucian concept of "*keisei saimin*" for production of high-quality medications.	Data not available	www.fujichemical.co.jp
	BioReal AB/ AstaReal, Sweden	Global producer of natural astaxanthin (AstaReal) and AstaReal Group, a member of the AstaReal family.	Data not available	www.astareal.se

Biomass and products derived from various marine microalgae, such as *Odontella*, *Pavlova*, and *Phaeodactylum*, are also marketed as nutraceuticals by a few small businesses due to the high concentration of essential LC-PUFA molecules, such as eicosapentaenoic acid (EPA), docosahexaenoic acid (DHA), or polysaccharides produced by *Porphyridium*. (Pulz and Gross, 2004). Microalgae type and pigment composition vary as chlorophylls, carotenoids, and phycobiliproteins. These pigments can be isolated from marine algae and demonstrate several biological, health, and food benefits. For example, lutein and β-carotene, carotenoids isolated from *Porphyra tenera*, have displayed suppressive effects alongside mutagen-induced *umu* C gene expression.

4.2.1 PHOTOSYNTHETIC BACTERIA

Cyanobacteria are a type of photosynthetic bacteria, consisting of >2,000 species split into 150 genera occupying open wetlands, ponds, ocean to mountain soils, and hot springs to snow fields. Most cyanobacteria are grown under abiotic stress effects on metabolites and are preferred for foods and feed applications. Some planktonic cyanobacteria, such as *Anabaena flos-aquae*, can change their buoyancy via specific gas vesicles in order to get the best possible light exposure. Other species excrete extracellular polysaccharides, which may account for up to 30% of their total polysaccharide content, resulting in the formation of floating aggregates (Pulz and Gross, 2004). In addition to their quick absorption and storage of nutrients, cyanobacteria commonly dominate algal communities. Phosphates are stored in polyphosphate granules; cyanophycin acts as a nitrogen, carbon, and energy reserve; and a highly branched -1,4-polyglucan molecule stores carbon and energy. Fission is the mode of reproduction, and sexual reproduction is not present. Cyanobacteria are accomplished for manufacturing a wide range of bioactive chemicals. When cyanobacterial blooms develop in bodies of water such as lakes, rivers, or drinking-water reservoirs, they may produce potent hepatotoxins or neurotoxins, which can create major consequences for public health (Pulz and Gross, 2004). Other secondary metabolites, such as antiviral agents, immunomodulators, inhibitors, and cytostatics, have the potential to be of therapeutic value (Leu and Boussiba, 2014). *Prochlorophyta* is another group of photosynthetic bacteria, belongs to phylum Cyanobacteria, responsible for unusual coloring justifies their isolation from other cyanobacteria species. For this reason, lack of phycobiliproteins, and the existence of chlorophyll a and b in *Prochlorophyta* are best characterized as "free-living chloroplasts." The color of these chloroplasts, as well as their thylakoid organization, is extremely like that of higher plant chloroplasts. *Prochloron*, *Prochlorothrix*, and *Prochlorococcus* are the only genera that are now recognized. Chlorophyta is generally classified in *Chlorophyceae*, *Trebouxiophyceae*, *Chlorodendrophyceae*, *Prasinophyceae*, *Dasycladophyceae*, and *Ulvophyceae*.

4.2.2 EUKARYOTIC MICROALGAE

Eukaryotic algae can be traced back around 1.9×10^9 years in phylogenetic history, making them much younger than cyanobacteria, which can be traced back approximately 2.7×10^9 years in phylogenetic history. Photosynthesis in plastids, which are long-lived organelles with green, brown, or blueish hues produced from endosymbiosis, occurs in eukaryotic algae, a varied group of unrelated species. Several evolutionary lineages formed from an endosymbiotic relationship between a cyanobacterium and the progenitor of the algae, including red and green algae, which seemed to have arisen at around the same time (Leu and Boussiba, 2014). For example, two primary families of green microalgae are identified based on biochemical and cellular distinctions: *Chlorophyta* and *Conjugaphyta* are a potential source of high-quality protein 40–65/100 g dry weight fit for human consumption. Despite being approximately five times bigger than *Chlorophyta*, none of the *Conjugaphyta* have yet to be used for biotechnological purposes. *Chlorophyceae* are the most numerous, with about 2,500 species divided into 350 genera. The majority of freshwater organisms are unicellular or filamentous. This category includes the most explored microalgae, such as *Chlorella*, *Chlamydomonas*, *Dunaliella*, and *Haematococcus*. Astaxanthin, a high-value coloring agent in aquaculture, particularly for trout and salmon, is another intriguing carotenoid. Efforts have been undertaken to cost-effectively generate astaxanthin from *H. pluvialis*, which accumulates up to 3% astaxanthin. Macroscopic algae make up the *Ulvophyceae* and *Charophyceae* families (Pulz and Gross, 2004). Although *Spirogyra* seems to generate bioactive chemicals, none of the unicellular or filamentous forms are biotechnologically important. Certain extremophilic algae develop at a slower pace than "ordinary" algae; this may be helpful in a variety of applications. A few thermophiles, such as *Mastigocladus laminosus* and *Cyanidium caldarium*, are species that thrive in diluted sulfuric acid (e.g., *D. acidophila*). *C. caldarium* and *Galdieria sulphuraria*, two red algae with strong growth rates at pH 1.0 and temperatures of 50°C, are both thermophilic and acidophilic. *G. sulphuraria* can also grow heterotrophically on roughly 50 distinct carbon sources, which no other microorganism can match. Salt tolerance and alkaliphily are commonly seen in species found in soda lakes (Leu and Boussiba, 2014).

4.3 BIOCHEMICAL COMPOSITION OF MARINE MICROALGAE

In addition to carbohydrates, proteins, lipids, nucleic acids, and other biologically important molecules, microalgae also produce vitamins and minerals (Yadavalli et al., 2021). According to algal strains and their physiological responses to environmental factors such as light intensity and photoperiod, temperature, and nutrition, the cellular makeup of each fraction varies. Table 4.2 shows the major biochemical components of selected microalgae species (Ryu et al., 2014). *Arthrospira* has been shown to have antiviral, anticancer, antioxidant, and anti-inflammatory properties, among other things. Green algae from the genus *Chlorella*, a single-celled, spherical microalga, have been promoted as containing a "growth factor," which is a water-soluble extract containing nucleic acids, amino acids, vitamins, minerals, polysaccharides, glycoproteins, and glucans (Figure 4.1). *Chlorella* is enriched with 11–18% protein, 12–18% carbohydrates, 2–46% lipids, and various amount of vitamins and amino acids. *Chlorella* extracts have been shown to have several therapeutic qualities, including cholesterol reduction, antioxidant, antibacterial, anticancer activity, and immune system enhancement (Ryu et al., 2014). Despite the stated advantages, multiple studies indicate that consuming large doses of *Chlorella* or *Arthrospira* may result in significant negative effects. Excessive use of *Chlorella*, for example, has been linked to allergies, nausea, vomiting, and other gastrointestinal issues. Some of these problems were observed as a result of the 2016 recall of Soylent products containing *Chlorella* algae flour as an ingredient. *Chlorella* has also been linked to acute tubulointerstitial nephritis, which may lead to renal failure. Diarrhea, nausea, and vomiting were all potential adverse effects of taking too much *Arthrospira*.

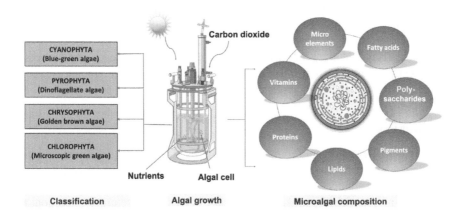

FIGURE 4.1 Classification of marine algae and biochemical composition.

TABLE 4.2
Microalgal Biocomponent Percentage in Eukaryotic Microalgae

Algal Species	Protein (w/w)	Carbohydrates (w/w)	Lipids (w/w)	References
Chlamydomonas reinhardtii	48%	17%	21.0%	(Becker, 2007)
Botryococcus braunii	40%	6%	33.1%	(Cheng et al., 2019)
Dunaliella salina	57%	32%	6.1%	(Becker, 2007)
Chlorella luteoviridis	47%	12%	22.1%	(José de Andrade and Maria de Andrade, 2017)
Chlorella sorokiniana	56%	17%	22.1%	(Chai et al., 2018)
Scenedesmus obliquus	56%	17%	14.2%	(Becker, 2007)
Haematococcus pluvialis	68%	9%	26.5%	(Olaizola, 2000)
Chlorella vulgaris	58%	17%	22.5%	(Becker, 2007)
Chaetoceros calcitrans	40%	37%	23.2%	(Velasco et al., 2017)

4.3.1 PROTEINS

Proteins are biological building blocks made up of various amino acids, which are crucial for all life forms on earth. The structure and metabolism of microalgae are heavily dependent on proteins. The membrane and light-harvesting complex, including multiple photosynthesis-related catalytic enzymes, depend on them. Many microalgal species have high protein content for both quantity and quality reasons compared to more traditional meat and fish sources (Graziani et al., 2013). This may vary from 42% to 70% in certain cyanobacteria and 58% for *Chlorella vulgaris* on a dry weight basis for protein concentrations. Microalgae contribute a variety of the necessary amino acids that humans cannot synthesize in their bodies, making them an excellent source of protein. There is also a similarity in the amino acid profiles between soy, lactoglobulin, and good egg albumin (Batista et al., 2013). Although microalgal protein application in food has been limited, since, microalgae-based foods include non-protein components (such the pigment chlorophyll), which may alter their color and flavor. Some strains of microalgae, such as *Chlorella* and *Tetraselmis*, have a stiff cell wall that reduces the extraction efficiency of intracellular proteins and results in poor digestibility in humans when the full biomass is employed. In the pharmaceutical and nutrition/food industries, peptides produced by microalgae have been proved to be active substitutes. The peptides utilized in medicinal applications are produced from the enzymatic breakdown of proteins and range in length from 2 to 30 amino acids. The peptides are inactive at first since they are part of the parent protein. The *Chlorella* sp. microalgae strand is one of the most widely utilized peptide makers (Hayes et al., 2017). Bioactive chemicals have received a lot of interest in the pharmaceutical sector because of their inherent safety, especially in anti-inflammatory, anticancer, antioxidant, and general human health applications.

4.3.2 CARBOHYDRATES

Carbohydrates are complicated organic compounds synthesized in microalgae during the photosynthesis process. In eukaryotic microalgal species, starch is predominantly synthesized from glucose and stored as amylose and amylopectin. Carbohydrates may be present in combination with protein or fat, and complex polysaccharides make up the bulk of cell wall structure. In addition, the primary carbon-containing photosynthetic products produced by microalgae are glucose and starch-like energy storage compounds. Microalgal carbohydrates consist primarily of starch, cellulose, and other polysaccharides. Phosphorus and nitrogen deficiency, like lipid deficiency, may promote carbohydrate buildup (Nwoba et al., 2016). *Chlorophyta* manufacture starch in the form of two glucose polymers, amylopectin and amylose, while *Rhodophyta* create floridean starch, a carbohydrate polymer. *Chrysolaminarin*, a linear polymer of (1,3) and (1,6) linked glucose units, is produced by diatoms (*Bacillariophycae, Heterokontophyta*). During the exponential development phase, certain diatoms may store up to 30% of their dry weight as (1,3)–D-glucan, and cells can accumulate up to 80% when nutrients are scarce. Similarly, under nutrient-rich vs. nutrient-depleted circumstances, a strain of *Tetraselmis suecica* was shown to accumulate between 11 and 47 % of its dry weight as starch (Graziani et al., 2013). Microalgal polysaccharides are widely used in food industries, preferably as gelling or thickening agents, and have also drawn attention in cosmetic industries for antioxidants for topical cream applications. A red alga, *Porphyridium cruentum*, a highly promising microalga known to produce sulphated galactan exopolysaccharide, has shown potential for carrageenans. Similarly, *Chlamydomonas mexicana* is an extracellular polysaccharide derived from microalga with pharmacological properties to stimulate the human immune system.

4.3.3 LIPIDS

Hydrocarbons, triacylglycerols, free fatty acids, phospholipids, sterols, glycolipids, and waxes are all types of lipids found in the lipid fraction. The amount and composition of microalgal lipids are mostly determined by classes, medium composition with a high C/N ratio and/or low nitrogen concentration, and varied stress situations. When the C/N ratio is high, nitrogenous resources are depleted more quickly, and metabolisms shift carbon flow from cell development to storage lipid synthesis (Chew et al., 2017). In all eukaryotic microalgae, these hydrocarbon organic compounds are often found inside the phylum of the organism. Polyunsaturated fatty acids have a higher economic significance as nutraceuticals and in baby formulas and provide health benefits for chronic illnesses. diabetes, arthritis, obesity, inflammatory disorders, and cardiovascular disease. A blend of C_{16} and C_{18} saturated and unsaturated fatty acids, as well as longer carbon-chain lengths, including several omega fatty acids, describes the fatty acid composition of microalgae (Williams and Laurens, 2010). These high-value products may be used in the food and pharmaceutical industries, among other places. However, from a medical standpoint, PUFAs have been shown to provide considerable protective impacts on human health and development. The lipid percentage of several algal species has been thoroughly characterized, and this proportion may range from 25% to 50% of the dry biomass (w/w). Nevertheless, different percentages reaching from 1% to 70% have also been recorded (Williams and Laurens, 2010).

4.3.4 VITAMINS

In aquatic ecosystems, microalgae act as a main food supplement and are accountable for the transmission of vital vitamins (e.g., A, B1, B2, B6, B12, C, E, nicotinate, biotin folic acid, and pantothenic acid) and minerals such as Na, K, Ca, Mg, Fe, and Zn through food chains. Because the human body does not have the capability to the produce vital vitamins and minerals on its own, the only way to receive them is via direct assimilation of raw microalgae or supplementation. Microalgae productivity and varied spectrum of high-value products are critical for the nutraceutical business. Vitamins A, B, C, D, and E are among the vitamins produced and accumulated by aquatic microalgae. High levels of vitamin B_{12} and Fe in spirulina make it suitable for nutritional supplements in baby formula. The amount and quality of the vitamins generated by microalgae are dependent on the genotype, growth cycle, harvesting period, light intensity, and growth conditions, according to a large body of research (Toti et al., 2018). The connections between optimum growth circumstances and vitamin synthesis vary greatly between species, according to the literature. However, vitamin production might vary even within the same species.

4.3.5 Polyphenols

Polyphenols are biopolymers of phenolic components that are members of a varied collection of 2° metabolites that have at least one hydroxyl functionality (-OH) connected to an aromatic unit. Moreover, polyphenols are polymers of phenolic compounds that are members of the phenolic chemical family. These compounds have features that are comparable to phytosterols; nevertheless, they are more efficient in radical scavenging than phytosterols are. Polyphenols offer a widespread range of biological characteristics, allowing them to be used extensively in the pharmaceutical and nutraceutical sectors, among other fields. While oxygen is necessary for life, it may have a deleterious influence on living organisms by producing oxidative stress in affected cells, which is a primary source of various malignancies in cancer, including breast cancer (Russell et al., 2022). These compounds act as biological antioxidants, which are of great relevance for treatment of different types of cancer, diabetes, and inflammation studies, among other fields.

4.3.6 Phytosterols

Tocopherols and sterols are extensively available in photosynthetic and non-photosynthetic networks of plants and microalga. Besides, phytosterols are a sub-class of steroids that are often indicated to be sterols in their own right. These compounds have a significant function in the pharmaceutical business, where microalgae are used in cholesterol-lowering, anti-inflammatory, antioxidant, and anticancer treatments, among other things. Sterols are extensively utilized to promote skin re-hydration and safeguard against free radicals, which is becoming more popular as natural skincare products expand in popularity (Volkman, 2016). Currently, the most important resource of phytosterols are vegetables and tall oils, which will have a market worth of $490 million by 2022, according to estimates. As an economical and well-established sector, the manufacture of phytosterols from vegetable oils is becoming more controversial due to rising worries about the long-term viability of the present production techniques, which require large amounts of land. Polyhydroxysterols from marine algae have been found to have anticancer, cytotoxic, and other significant biological activities.

4.3.7 Carotenoids

Carotenoids are natural color pigments and mainly consist of 18-carbon conjugated structures with two hexa carbon rings at each end. Carotenoids are high-value compounds essential during de novo designed light-activated Reaction Centers by photosynthetic organisms and provide photoprotection through the absorption of light within chlorophyll (Singh et al., 2022). In general, these color pigments play a significant role in shielding microalgae from free radical outbreak while quenching the single oxygen and trapaperxoyl radicals. Carotenoids are mostly preferred in cosmetic industries due to their antioxidant properties and are traditionally used for natural coloring of food, among other things. *Dunaliella salina* is a unicellular microalga with 98.5% of β-carotene, glycerol, and protein, which corresponds to 13% of dry biomass. β-carotene has demonstrated the capability to reduce cancer cells' growth by inducing apoptosis in an extensive range of cancer cells and have broad application in pharmaceutical and nutraceuticalsindustries. Betatene, Western Biotechnology, AquaCarotene Ltd., Cyanotech Corp., and Nature Beta Technologies, located in the United States, China, and Australia, are commercially producing *Dunaliella* carotenoids. Astaxanthin is another frequently commercialized carotenoid, well known for its distinctive red pigment preferred for food dye, antioxidant properties, and free radical absorption capacity. *Haematococcus pluvalis*, a unicellular biflagellate green alga, has been shown to accumulate 81% of carotenoids corresponding to 7% weight; prevent oxidative damage; be beneficial for the treatment of diabetes and neurodegenerative disorders; and stimulate immunization.

4.4 MICROALGAE AS FOOD SUPPLEMENT

Biologically active compounds extracted from microalgae, like protein, polyunsaturated fatty acids, pigments, vitamins, minerals, and fibers, are regarded as favorable food supplements for human consumption attributed to their sensible cellular configuration (Singh et al., 2022). Nowadays, microalgae are industrially cultivated as a foundation for high added value chemical precursors. For instance, long PFAs are indeed able to produce omega 3-fatty acids, including linoleic acid, γ-linolenic acid, and arachidonic acid (Camacho et al., 2019). Eicosapentaenoic acid and docosahexaenoic acid have been known to reduce cardiovascular effusions, arthritis, and hypertension and exhibit hypolipidemic activity. Several microalgae species, such as (Chlorophyceae) *Chlorella vulgaris, Haematococcus pluvialis, Dunaliella salina*, and the cyanobacteria spirulina, have been industrially grown worldwide to provide nutritional supplements for human food. For example, *Chlorella vulgaris* and *Arthrospira plantesis* have been recently employed as alternative medicine and serve as a protein source compared to dairy protein in the nutrition, cosmetic, and food industries and are widely accepted for treatment of gastric ulcers, wounds, constipation, anemia, hypertension, diabetes infant malnutrition, and so on. Japan and Korea are industrially cultivating *Chlorella* sp. microalgae for food additives. Several studies have shown that carbohydrates exhibit prebiotic properties desirable for virucidal, antibacterial, antifungal, anti-inflammatory, immunomodulatory, anticoagulant/antithrombotic, antiproliferative/tumor suppressor, antilipidemic, and hypoglycemic agent applications. *Haematococcus pluvialis*, a carotenoid pigment, has been known for high astaxanthin content from 1.5–3.0% on oven dry weight and serves as a potential radical scavenger and singlet oxygen quencher.

4.5 BIOLOGICAL ACTIVITY FROM MICROALGAE

4.5.1 Anticancer Activity

Free radicals are responsible for the development of cancer-causing cells, which go through the stages of initiation, promotion, and advancement the same way normal cells do. Many studies have been performed on natural pigments such as chlorophyll, carotenoids, and derivatives of these substances to determine their usefulness as cancer chemo-preventative agents (Hafsa et al., 2017). The antimutagenic and antitumorigenic properties of these chemicals also contribute to suppression of the activity of cancer cells during their development (Figure 4.2). A cytotoxicity experiment was performed to assess the anticancer activity of pigments, and it was discovered that anticancer activity was greater under nitrogen-strained circumstances than under non-nitrogen-stressed ones. As a result of this, natural pigments derived from microalgae may have the potential to serve as chemo-preventive reagents for the prevention of carcinogenesis (Table 4.3) (Hafsa et al., 2017). For example, fucoxanthin is the most abundant carotenoid, contributing to 10% of total carotenoid production, and has been investigated for anticancer activity for the treatment of oxidative destruction, responsible for carcinogenesis (Ishikawa et al., 2008). Ishikawa et al. (2008) investigated the inhibitory effect of anti-adult T-cell leukemia effects of fucoxanthin and its deacetylated intermediate chemical compound, fucoxanthinol.

4.5.2 Anti-Inflammatory Activity

Polyunsaturated fatty acids and pigments from microalgae have shown anti-inflammatory properties, indicating the possibility for dietary components to treat chronic inflammatory illnesses (Robertson et al., 2015). Various *in vivo* investigations confirmed that microalgae components like carotenoids have antioxidant and anti-inflammatory properties in clinical trials (Deng and Chow, 2010). This confirms microalgae's capacity to cure inflammatory and other disorders. Fucoxanthin, a natural carotenoid in diatom brown algae and crypto algae, exhibited 5,6-monocyclic oxidation

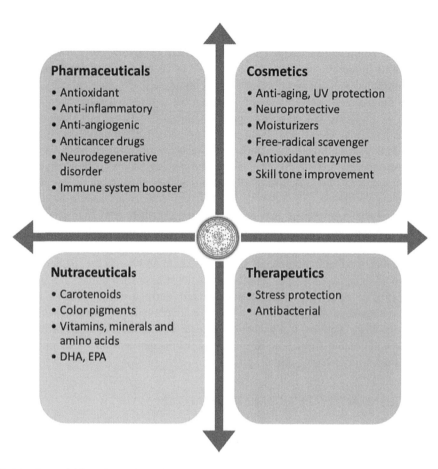

FIGURE 4.2 Potential benefits and applications of marine algae.

TABLE 4.3
Microalgae Bioactive Compounds and Their Health Benefits (Zhuang et al., 2022)

Microalgae	Isolated Compounds	Biological Application
Phaeodactylum	Eicosapentaenoic acid	Antioxidant, anti-inflammatory and Nutraceutical
Arthrospira	Linolenic acid	Immunity booster, tissue repair and anti-ageing characteristics
Pavlova	Brassica sterol	Reduces hyper cholesterol
Arthrospira	Phycocyanin	Antioxidant, anti-inflammatory and Nutraceutical
Dunaliella	Beta-carotene	Health promotion, anticancer, coronary diseases prevention, UV protection and antioxidant
Haematococcus	Astaxanthin	Anti-ageing, antioxidant, and age-macular protection
Porphyridium	Superoxide dismutase	Antioxidant, anti-inflammatory
Tetraselmis	Alpha-tocopherol	Antioxidant

TABLE 4.3 (CONTINUED)

Microalgae	Isolated Compounds	Biological Application
Porphyridium	Aminobutyric acid	CNS improvement and Nutraceutical
Gambierdiscus	Okadaic	Antifungal and anti-inflammatory
Chlorella, Skeletonema	Polysaccharides, glucan, Lutein, oleic acid, canthaxanthin	Moisturizing and thickener agent, antimicrobial and antibiofilm
Arthrospira platensis	Exopolysaccharides	Antioxidant and anti-inflammatory
Haematococcus pluvialis	Astaxanthin	Antioxidant, UV protection
Spirulina, *Porphyridium*	Phycocyanobilin phycoerythrobilin, polyunsaturated fatty acids, diacylglycerols, phycocyanin	Antioxidant and color pigmentation and food application
Nostoc linckia	Borophycin, cryptophycin	Anti-tumor compounds

and allene bonds. Fucoxanthin has anti-hypertensive, anticancer, and anti-inflammatory effects and is inhibitory to cancer cells.

4.5.3 ANTI-ANGIOGENIC ACTIVITY

Angiogenesis is the establishment of a vascular arrangement of new blood vessels, which are necessary for organ development and repair in different pathologic conditions such as tumor growth and metastasis. Under pathological circumstances, uncontrolled angiogenesis is likely to promote disease transmission. As a result, angiogenesis inhibitors are necessary, since they provide more effective therapy with fewer adverse effects. Bioactive fractions derived from brown marine alga, that is, *Chnoospora minima* extracts, contain (carbon tetrafluoride) CF4 bioactive compounds and could efficiently suppress the growth of cancer on chick embryos. Fucoxanthin, a bioactive compound extracted from brown marine algae, has shown anti-angiogenic activity and could prevent angiogenesis-related diseases and diabetic retinopathy.

4.5.4 ANTIOXIDANT ACTIVITY

Microalgae have natural antioxidant properties suitable for cosmetics, sunscreen, and food/nutraceutical preparations (Table 4.3). Several reports documents that methanolic microalgal crude extracts derived from *Isochrysis galbana, Chlorella vulgaris, Nannochloropsis oculata, Tetraselmis tetrathele,* and *Chaetoceros calcitrans* have an enriched proportion of tocopherol. β-carotene loaded on micelles has been investigated to have higher antioxidant properties for cosmetic application (Gui et al., 2022). Chlorella extract has a strong affinity for skin and hair and reduces wrinkles, making skin radiant; thus chlorella extract can be used as an antioxidant and regulator within sun protection creams. *Dunaliella salina* is capable of accumulating high β-carotene (14% of the dry cell weight) in eukaryote microalgal cells and imparting red color to water bodies. Spirulina, such as astaxanthin, is a known effective antioxidant to improve human metabolism, helpful for cell proliferation and inhibiting the cell cycle through induction of cell apoptosis. Certain microalgal polysaccharides such as *Nostoc commune* have also shown antioxidant activity and moisturizing characteristics *in vitro* and *in vivo* suitable for cosmetic preparations.

4.5.5 NEUROPROTECTIVE ACTIVITY

Microalgal-derived bioactive compounds and cell extracts have demonstrated antioxidant activities, inhibition of cholinesterases, oxygen/glucose deprivation, hydrogen peroxide, glutamate, and prevention of β-amyloid aggregation and neuronal damage. Certain microalgae-derived carotenoids, PUFAs, phenolic compounds, and sterols exhibit neuroprotective activity of microalgae against Alzheimer's disease via anticipation of nerve cell injury and inhibition of radical-induced neuronal destruction (Olasehinde et al., 2017). These bioactive compounds received from algal species can serve as neuroprotective representatives while chelating with copper, iron, and calcium. *Chlorella vulgaris*, a microalgal extract, could attenuate oxidative stress and enhanced cell death. Similarly, astaxanthin, a carotenoid extracted from microalgae, displayed neuroprotective properties via limiting hydrogen peroxide-induced neuronal destruction through proliferation of protein kinase B (Akt) and extracellular signal-regulated protein kinase-1/2 (ERK1/2) metabolic pathways. Microalgal polyunsaturated fatty acids inhibit uncharacteristic synaptic transmission and neuronic death by modulation of cholinergic properties and enhancing nerve impulse transmission.

4.6 SUMMARY AND CONCLUSIONS

Despite the difficulties associated with industrial growth and commercial-scale production, there is an urgent need for development of microalgal product applications in meals, supplements, and possible medications. Nevertheless, considerable further study is required in this sector to define the biochemical composition of prospective algae to fully comprehend their advantages and any potential risks they may pose. Researchers have identified microalgal protein hydrolysates and isolated peptides to have the potential to be employed as antioxidants and antihypertensive and anticancer agents. However, only a few commercially viable microalgae production strains have been investigated yet (in comparison to more than 100,000 predicted strains). There are tremendous opportunities for the development and synthesis of novel bioactive compounds exhibiting potential antimicrobial, antifungal, and antitumorigenic (as well as other) characteristics. Few existing microalgae extracts have been explored in the pharmaceutical and cosmetic industries; thus it is necessary to overcome the difficulty of technology applications for large-scale algal biomass production. Employment of advanced genetic engineering tools, strain improvement, high biomass productivity, and bioactive compound extraction should integrate with digital control of the microalgae industry to prepare the marine algae industry for a better future.

REFERENCES

Barkia, I., Saari, N., Manning, S.R., 2019. Microalgae for high-value products towards human health and nutrition. Mar. Drugs 17, 304. https://doi.org/10.3390/md17050304

Batista, A.P., Gouveia, L., Bandarra, N.M., Franco, J.M., Raymundo, A., 2013. Comparison of microalgal biomass profiles as novel functional ingredient for food products. Algal Res. 2, 164–173. https://doi.org/10.1016/j.algal.2013.01.004

Becker, E.W., 2007. Micro-algae as a source of protein. Biotechnol. Adv. 25, 207–210. https://doi.org/10.1016/j.biotechadv.2006.11.002

Borowitzka, M.A., 2013. High-value products from microalgae-their development and commercialisation. J. Appl. Phycol. 25, 743–756. https://doi.org/10.1007/s10811-013-9983-9

Camacho, F., Macedo, A., Malcata, F., 2019. Potential industrial applications and commercialization of microalgae in the functional food and feed industries: A short review. Mar. Drugs 17. https://doi.org/10.3390/md17060312

Chai, S., Shi, J., Huang, T., Guo, Y., Wei, J., Guo, M., Li, L., Dou, S., Liu, L., Liu, G., 2018. Characterization of *Chlorella sorokiniana* growth properties in monosaccharide-supplemented batch culture. PLoS One 13, e0199873. https://doi.org/10.1371/journal.pone.0199873

Cheng, P., Okada, S., Zhou, C., Chen, P., Huo, S., Li, K., Addy, M., Yan, X., Ruan, R.R., 2019. High-value chemicals from *Botryococcus braunii* and their current applications—a review. Bioresour. Technol. 291, 121911. https://doi.org/10.1016/j.biortech.2019.121911

Chew, K.W., Yap, J.Y., Show, P.L., Suan, N.H., Juan, J.C., Ling, T.C., Lee, D.-J., Chang, J.-S., 2017. Microalgae biorefinery: High value products perspectives. Bioresour. Technol. 229, 53–62. https://doi.org/10.1016/j.biortech.2017.01.006

Deng, R., Chow, T.-J., 2010. Hypolipidemic, antioxidant, and antiinflammatory activities of microalgae spirulina. Cardiovasc. Ther. 28, e33–e45. https://doi.org/10.1111/j.1755-5922.2010.00200.x

Eltanahy, E., Torky, A., 2021. Chapter 1: Microalgae as cell factories: Food and feed-grade high-value metabolites, in: Microalgal Biotechnology: Recent Advances, Market Potential, and Sustainability. The Royal Society of Chemistry, pp. 1–35. https://doi.org/10.1039/9781839162473-00001

Enamala, M.K., Chavali, M., Pamanji, S.R., Tangellapally, A., Dixit, R., Singh, M., Kuppam, C., 2021. Greener synthesis of enzymes from marine microbes using nanomaterials, in: Handbook of Greener Synthesis of Nanomaterials and Compounds. Elsevier, pp. 109–136. https://doi.org/10.1016/B978-0-12-822446-5.00005-8

Graziani, G., Schiavo, S., Nicolai, M.A., Buono, S., Fogliano, V., Pinto, G., Pollio, A., 2013. Microalgae as human food: Chemical and nutritional characteristics of the thermo-acidophilic microalga *Galdieria sulphuraria*. Food Funct. 4, 144–152. https://doi.org/10.1039/C2FO30198A

Gui, L., Xu, L., Liu, Z., Zhou, Z., Sun, Z., 2022. Carotenoid-rich microalgae promote growth and health conditions of *Artemia nauplii*. Aquaculture 546, 737289. https://doi.org/10.1016/j.aquaculture.2021.737289

Hafsa, M.B., Ismail, M.B., Garrab, M., Aly, R., Gagnon, J., Naghmouchi, K., 2017. Antimicrobial, antioxidant, cytotoxic and anticholinesterase activities of water-soluble polysaccharides extracted from microalgae *Isochrysis galbana* and *Nannochloropsis oculata*. J. Serbian Chem. Soc. 82, 509–522. https://doi.org/10.2298/JSC161016036B

Hayes, M., Skomedal, H., Skjånes, K., Mazur-Marzec, H., Toruńska-Sitarz, A., Catala, M., Isleten Hosoglu, M., García-Vaquero, M., 2017. Microalgal proteins for feed, food and health, in: Microalgae-Based Biofuels and Bioproducts. Elsevier, pp. 347–368. https://doi.org/10.1016/B978-0-08-101023-5.00015-7

Hussain, F., Shah, S.Z., Ahmad, H., Abubshait, S.A., Abubshait, H.A., Laref, A., Manikandan, A., Kusuma, H.S., Iqbal, M., 2021. Microalgae an ecofriendly and sustainable wastewater treatment option: Biomass application in biofuel and bio-fertilizer production. A review. Renew. Sustain. Energy Rev. 137, 110603. https://doi.org/10.1016/j.rser.2020.110603

Ishikawa, C., Tafuku, S., Kadekaru, T., Sawada, S., Tomita, M., Okudaira, T., Nakazato, T., Toda, T., Uchihara, J.-N., Taira, N., Ohshiro, K., Yasumoto, T., Ohta, T., Mori, N., 2008. Antiadult T-cell leukemia effects of brown algae fucoxanthin and its deacetylated product, fucoxanthinol. Int. J. Cancer 123, 2702–2712. https://doi.org/https://doi.org/10.1002/ijc.23860

José de Andrade, C., Maria de Andrade, L., 2017. An overview on the application of genus *Chlorella* in biotechnological processes. J. Adv. Res. Biotechnol. 2, 1–9. https://doi.org/10.15226/2475-4714/2/1/00117

Kholssi, R., Ramos, P.V., Marks, E.A.N., Montero, O., Rad, C., 2021. Biotechnological uses of microalgae: A review on the state of the art and challenges for the circular economy. Biocatal. Agric. Biotechnol. 36, 102114. https://doi.org/10.1016/j.bcab.2021.102114

Leu, S., Boussiba, S., 2014. Advances in the production of high-value products by microalgae. Ind. Biotechnol. 10, 169–183. https://doi.org/10.1089/ind.2013.0039

Nwoba, E.G., Ayre, J.M., Moheimani, N.R., Ubi, B.E., Ogbonna, J.C., 2016. Growth comparison of microalgae in tubular photobioreactor and open pond for treating anaerobic digestion piggery effluent. Algal Res. 17, 268–276. https://doi.org/10.1016/j.algal.2016.05.022

Olaizola, M., 2000. Commercial production of astaxanthin from *Haematococcus pluvialis* using 25,000-liter outdoor photobioreactors. J. Appl. Phycol., pp. 499–506. https://doi.org/10.1023/a:1008159127672

Olasehinde, T.A., Olaniran, A.O., Okoh, A.I., Koulen, P., 2017. Therapeutic potentials of microalgae in the treatment of Alzheimer's disease. Molecules 22, 1–18. https://doi.org/10.3390/molecules22030480

Pulz, O., Gross, W., 2004. Valuable products from biotechnology of microalgae. Appl. Microbiol. Biotechnol. 65, 635–648. https://doi.org/10.1007/s00253-004-1647-x

Russell, C., Rodriguez, C., Yaseen, M., 2022. High-value biochemical products & applications of freshwater eukaryotic microalgae. Sci. Total Environ. 809, 151111. https://doi.org/10.1016/j.scitotenv.2021.151111

Ryu, N.H., Lim, Y., Park, J.E., Kim, J., Kim, J.Y., Kwon, S.W., Kwon, O., 2014. Impact of daily *Chlorella* consumption on serum lipid and carotenoid profiles in mildly hypercholesterolemic adults: A double-blinded, randomized, placebo-controlled study. Nutr. J. 13, 57. https://doi.org/10.1186/1475-2891-13-57

Singh, M., Chavali, M., Enamala, M.K., Obulisamy, P.K., Dixit, R., Kuppam, C., 2020. Algal Bioeconomy: A Platform for Clean Energy and Fuel. Springer, pp. 335–370. https://doi.org/10.1007/978-981-15-9593-6_13

Singh, M., Mal, N., Mohapatra, R., Bagchi, T., Parambath, S.D., Chavali, M., Rao, K.M., Ramanaiah, S.V., Kadier, A., Kumar, G., Chandrasekhar, K., Kim, S.-H., 2022. Recent biotechnological developments in reshaping the microalgal genome: A signal for green recovery in biorefinery practices. Chemosphere 293, 133513. https://doi.org/10.1016/j.chemosphere.2022.133513

Toti, E., Chen, C.-Y.O., Palmery, M., Villaño Valencia, D., Peluso, I., 2018. Non-provitamin A and provitamin A carotenoids as immunomodulators: Recommended dietary allowance, therapeutic index, or personalized nutrition? Oxid. Med. Cell. Longev. 2018, 1–20. https://doi.org/10.1155/2018/4637861

Velasco, L.A., Carrera, S., Barros, J., 2017. Isolation, culture and evaluation of *Chaetoceros muelleri* from the Caribbean as food for the native scallops, *Argopecten nucleus* and *Nodipecten nodosus*. Lat. Am. J. Aquat. Res. 44, 557–568. https://doi.org/10.3856/vol44-issue3-fulltext-14

Volkman, J.K., 2016. Sterols in microalgae, in: The Physiology of Microalgae. Springer International Publishing, pp. 485–505. https://doi.org/10.1007/978-3-319-24945-2_19

Williams, P.J. L. B., Laurens, L.M.L., 2010. Microalgae as biodiesel & biomass feedstocks: Review & analysis of the biochemistry, energetics & economics. Energy Environ. Sci. 3, 554. https://doi.org/10.1039/b924978h

Yadavalli, R., Ratnapuram, H., Motamarry, S., Reddy, C.N., Ashokkumar, V., Kuppam, C., 2020. Simultaneous production of flavonoids and lipids from *Chlorella vulgaris* and *Chlorella pyrenoidosa*. Biomass Convers. Biorefinery, 1–9. https://doi.org/10.1007/s13399-020-01044-x

Yadavalli, R., Ratnapuram, H., Peasari, J.R., Reddy, C.N., Ashokkumar, V., Kuppam, C., 2021. Simultaneous production of astaxanthin and lipids from *Chlorella sorokiniana* in the presence of reactive oxygen species: A biorefinery approach. Biomass Convers. Biorefinery, 1–9. https://doi.org/10.1007/s13399-021-01276-5

Zhuang, D., He, N., Khoo, K.S., Ng, E.-P., Chew, K.W., Ling, T.C., 2022. Application progress of bioactive compounds in microalgae on pharmaceutical and cosmetics. Chemosphere 291, 132932. https://doi.org/https://doi.org/10.1016/j.chemosphere.2021.132932

5 Filler Feed from Marine Micro- and Macroalgae

Dillirani Nagarajan, Duu-Jong Lee and Jo-Shu Chang

CONTENTS

5.1	Introduction	61
5.2	Aquaculture Feed Composition and Its Significance	62
5.3	Marine Microalgae as a Potential Filler Feed Ingredient	63
	5.3.1 Carbohydrates	63
	5.3.2 Lipids	64
	5.3.3 Pigments and Antioxidants	64
5.4	Marine Macroalgae as a Potential Filler Feed Ingredient	67
5.5	Conclusions and Perspectives	72
References		72

5.1 INTRODUCTION

Global aquaculture production is expected to increase from 81.2 million tonnes in 2018 to 109 million tonnes by 2030, accounting for 53% of the total global fisheries production by 2030 (Food and Agriculture Organization of the United Nations, 2020). This tremendous increase in aquaculture is placing an enormous strain on the capture fisheries sector since fishmeal and fish oil are still considered the most appropriate feed for aquaculture. For the production of 1 kg of meat, 1.1–1.6 kg of feed is required, and the feed must meet the regulatory standards required for feed applications. The feed must provide all the essential nutrients (in the form of proteins, vitamins, minerals, and lipids), maintain the growth and survival rates of the farmed fish, not interfere with the meat quality, and be sustainable with minimal environmental impacts (Jones et al., 2020). About 18.3% of capture fisheries are used in aquaculture as feed ingredients (Boyd et al., 2022). Sustainable fishing and conservation of marine diversity to prevent the overfishing of certain trophic fish are part of the sustainable development goals developed by the United Nations. Thus, alternative aquaculture feed ingredients are needed for sustainable aquaculture.

Algae, including micro-and macroalgae, are photosynthetic organisms, capable of converting light energy into chemical energy in the form of organic biomass. Microalgae have been recognized as a third-generation biofuel feedstock due to high aerial productivity, non-requirement of agricultural land, high photosynthetic efficiency, and superior adaptability to harsh environmental conditions. Microalgae are rich in organic nutrients such as proteins, lipids, and carbohydrates. In addition, functional and bioactive compounds, including polyunsaturated fatty acids (PUFAs), pigments, and carotenoids (chlorophyll, astaxanthin, carotenes), in microalgae confer potent anti-microbial, antioxidant, anticancer, anti-inflammatory, and UV protection properties. Aquaculture feeds are typically rich in proteins, and microalgae are often dubbed single-cell proteins for their applications as protein-rich health supplements. Antioxidant pigments and PUFAs in microalgae confer higher stress tolerance and improved fatty acid profile in farmed animals. Microalgae have been traditionally used as aquaculture feed, particularly in larval hatcheries for fish and shrimps (Ranglová et al., 2022). Of the global microalgal biomass production of 22,000–25,000 tons, 30% is procured for feed applications.

Macroalgae, commonly called seaweeds, are benthic aquatic plant-like organisms frequently found in marine habitats. Macroalgae are a diverse group of organisms, and they are divided into

DOI: 10.1201/9781003326946-5

three major divisions based on their pigmentation—*Chlorophyta* (green macroalgae), *Rhodophyta* (red macroalgae), and *Phaeophyta* (brown macroalgae). Of these, red and brown macroalgae are extensively farmed in onshore and offshore farms for their potential use in the production of hydrocolloids. Macroalgal polysaccharides such as agar, carrageenan, alginate, and ulvan have superior gelling properties and are extensively used as thickening agents in the food, pharmaceutical, and cosmetic industries. Furthermore, macroalgae are a treasure house of bioactive poly- and oligosaccharides with immune-modulating properties. Feeding small quantities of macroalgal meal has been shown to improve immune response, oxidative stress response, anti-inflammatory, and anti-cytotoxic activities in experimental model zebrafish (Hoseinifar et al., 2022). In recent times, macroalgal meal is gaining attention as an animal and aquaculture feed ingredient. Commercially available "Oceanfeed" is a mix of seaweeds that can be used as a direct feed for bovines, sows, aquaculture, and poultry (Satessa et al., 2020).

This chapter discusses the role of marine microalgae and macroalgae as suitable filler feeds in aquaculture. Filler is an ingredient often used to bulk up feed quantity without significantly altering the nutritional or calorific content of the intended feed. In general, non-digestible fiber or complex carbohydrates are used as fillers in feed, which do not interfere with the calorific value but aid in gastric emptying and general metabolic health (Müller et al., 2018). Both microalgae and macroalgae are processed in industrial settings for oil or polysaccharide extraction, generating a spent algal meal, which has potential applications as filler feed. The role of the algal biomass components in the feed is discussed in detail in the subsequent sections.

5.2 AQUACULTURE FEED COMPOSITION AND ITS SIGNIFICANCE

Typically, an aquaculture feed consists of proteins, carbohydrates, lipids, vitamins, minerals, and other supplements, such as prophylactic antibiotics, supportive growth hormones, and antioxidants. Aquaculture feeds are usually protein rich since most of the fish are carnivorous with high protein requirements. Proteins act as a source of amino acids which support biomass growth, metabolic health, and the overall well-being of the farmed fish. Lipids are essential nutrients for farmed fish, as they are an additional source of energy. They aid in the absorption of fat-soluble vitamins, improve the palatability of feed, and provide fatty acids required for membrane lipid development and maintenance. Carbohydrates are considered an energy source and provide the calorific needs of feed. Carbohydrate addition to feeds is carefully modulated not to interfere with the nutritional content of feeds. Vitamins and minerals are essential cofactors for various enzymatic reactions, and deficiency results in disease outbreaks in aquaculture. Farmed aquatic animals are subjected to a lot of environmental stress, particularly due to the intense farming habits practiced currently. So additives such as antioxidants and antibiotics are essential to alleviate oxidative stress and pathogen infection, respectively. In addition, a binder is used to bind together all the ingredients in the feeds, and commonly used binders are wheat gluten and agar gum. Based on the purpose, feeds in aquaculture are of different types: whole/complete feeds, which take into account the nutritional and energy requirements of the farmed fish; supplemental feeds that provide a specific component such as polyunsaturated fatty acids or pigments to fortify the fish; a concentrated feed that can be further diluted before feeding; and a pre-mix made of specific ingredients that are completely soluble (Suresh, 2016).

Fillers are non-digestible or inert bulking agents used for improving the physical properties of the feed pellets. Fillers are used about 10–20% by weight maximum in feeds. A critical characteristic of the filler is that it should not interfere with nutrient digestibility and availability, subsequently growth, metabolic health, or overall wellbeing of the farmed aquatic species. In other words, feeds should remain isoenergetic and isonitrogenous even after the addition of fillers. Most often, non-digestible fibers such as cellulose and carboxymethyl cellulose are used as fillers in aquaculture feed. Rainbow trout adapted to 30% α-cellulose as fillers in feed by increasing the feed consumption to compensate for the decrease in available metabolic energy and increased gastric emptying

(Bromley & Adkins, 1984). Sometimes fillers might affect feed conversion efficiencies (feed consumed vs wet weight obtained). This reduction is not attributed to the negative effects of the filler used but to the reduction in the available nutrients due to the addition of fillers. A 10% addition of starch, cellulose, or a natural zeolite did not interfere with the nutrient digestibility, protein retention, or growth performance of seabass. However, an increase to 20% reduced feed efficiency, mainly due to the dilution of nutrients, in addition to increased gastric emptying (Dias et al., 1998). Conversely, protein fractions rendered inert by formaldehyde treatment have been used as a non-digestible filler to optimize the dietary protein requirements of certain fish (Barreto-Curiel et al., 2019; Durazo et al., 2010). The use of marine microalgae and macroalgae might not interfere with nutrient availability and digestibility of the feed but confer numerous positive health effects due to their repertoire of bioactive compounds such as pigments, antioxidants, and polyunsaturated fatty acids.

5.3 MARINE MICROALGAE AS A POTENTIAL FILLER FEED INGREDIENT

The proximate biochemical composition of microalgae biomass includes carbohydrates, proteins, lipids, and additional secondary compounds with specific physiological functions such as photosynthetic pigments, including carotenoids, polyunsaturated fatty acids, and sterols. Of all the nutrients, microalgal biomass has been explored as a replacement for fishmeal as a protein source, and these aspects were discussed previously (Kusmayadi et al., 2021). The key function of each of these ingredients in aquaculture feed is as follows (Nagarajan et al., 2021):

i. *Carbohydrates*—primary energy source for metabolizable energy and contribute to the net energy content of the feed.
ii. *Proteins*—crucial nutrient in aquaculture feeds, an essential nutrient in the form of amino acids required for biomass growth, vital for increasing protein content in the fillet.
iii. *Lipids*—the source of metabolizable energy.
iv. *Polyunsaturated fatty acids (PUFA)*—enhancement of the PUFA content of fish, improved lipid and fat profile of the final fillet.
v. *Pigments and carotenoids*—protect against oxidative stress by acting as potent antioxidants and improving the color palate of salmonid fish and shrimps and the PUFA content of the fillet.
vi. *Dietary fiber*—improved gastric emptying, better blood profile for serum lipids and glucose, and aid in flushing of toxins from the animals.

5.3.1 Carbohydrates

Microalgal carbohydrates are a part of the cell wall as structural polysaccharides or serve as energy reserves in the cells as storage reserves. Starch, glycogen, and other metabolizable polysaccharides serve as the energy source, while the resistant polysaccharides constitute dietary fibers.

The non-digestible carbohydrates present in the cell wall of microalgal cells serve as dietary fibers and form a key filler feed ingredient. Microalgae are phenotypically diverse and contain cell wall polymers composed of mannans, fructans, uronic acids, and rhamnans with high structural diversity, including cellulose, hemicellulose, and very small quantities of other polysaccharides. The cell wall of the most commonly used single-cell protein *Chlorella* is known to have chitin-like glucosaminoglycan in addition to galactans, rhamnans, and cellulose (Kapaun & Reisser, 1995). Animal intestines are devoid of complex carbohydrate hydrolyzing enzymes; thus the insoluble/indigestible cell wall polysaccharides act as dietary fiber. The processing of microalgal biomass for protein or pigment extraction results in a fiber-rich post-extraction residue, because the cell wall components are retained in the residue. The dietary fiber content of lipid-extracted *P. tricornutum* biomass consisted of 43.54% of insoluble dietary fiber (German-Báez et al., 2017). *Tetraselmis suecica*, a common aquaculture feed alga, was shown to contain 14–17% of dietary fiber in the biomass composition, similar to *P. tricornutum* and *Porphyridium purpureum* (Niccolai et al., 2019). These complex polysaccharides are fermented in the gut of the receiving animal, acting as a prebiotic in some cases. On the other hand, they function as dietary fibers aiding in gastric emptying, as mentioned previously.

In addition, cellular polysaccharides show potential health benefits. Microalgae storage polysaccharides are diverse, and certain unique polysaccharides possess supportive biological activity. *Parachlorella kessleri* HY1 cell wall extract revealed the existence of xylogalactofuranan and rhamnan as constituents. The polysaccharides exhibited cytotoxic activity and pro-inflammatory response *in vitro* (Sushytskyi et al., 2020). The common aquaculture feed alga *Isochrysis galbana* contains various soluble and insoluble polysaccharides, and a glucan with α(1→6) linked glucose backbone with branches showed potential anti-tumor activity against human lymphoma cells (Sadovskaya et al., 2014). Similarly, a novel β-type heteropolysaccharide composed of rhamnose, glucose, galactose, and mannose isolated from *I. galbana* showed antioxidant activities against hydroxyl and superoxide radicals (Sun et al., 2014).

5.3.2 Lipids

Microalgae have been overexploited and overexplored, especially for lipid content when they were looked upon as a potential source of fatty acids for biofuel production. Microalgal lipids are of three types: structural lipids—phospholipid, glycolipids, and betaines present in cell membranes and organelle membranes; storage lipids—glycerolipids or triacylglycerols used for the production of biodiesel; and functional lipids—polyunsaturated fatty acids and sterols with beneficial health effects (Pina-Pérez et al., 2019). Typically, microalgae rich in PUFAs have been used as aquaculture feeds to improve the PUFA content of oily fish such as salmonids. PUFAs are neutral fatty acids with an unsaturated backbone, and a double bond at the 3rd, 6th, or 9th carbon is denoted as omega-3, omega-6, and omega-9 fatty acid, respectively. Animals cannot synthesize some fatty acids, and these are called essential fatty acids and must be supplemented in the diet. Essential fatty acids for aquaculture feed include docosahexaenoic acid, γ-linolenic acid (GLA), linoleic acid (LA), α-linolenic acid (ALA), arachidonic acid (AA), and eicosapentaenoic acid (Parrish, 2009). PUFA-rich microalgae can replace fish oils in aquaculture feeds. Commercially sourced DHA-rich algal oils were used in combination with rapeseed oil and poultry oil to replace fish oil in the diets of gilthead sea bream *Sparus aurata*. This combination could completely replace 15% fish oil in feed without any alterations in the fatty acid profile of the fish. (Carvalho et al., 2020). Inclusion of *Nannochloropsis oceanica* (10%) in the feeds of the Atlantic salmon could replace 50% fish meal and 10% fish oil without any significant changes in the growth, feed efficiency, fillet fatty acid profile, or intestine morphology (Gong et al., 2020). Whole cells of the DHA-rich thraustochytrid *Schizochytrium* sp. could serve as aquaculture feed for *Salmo salar* without any pretreatment or oil extraction process, with 98% digestible and bioavailable fatty acids and proteins (Hart et al., 2021).

Furthermore, microalgae consist of phytosterols such as β-sitosterol (24-α-ethyl cholesterol), stigmasterol (Δ22,24-α-ethyl cholesterol), and campesterol (24-α-methyl cholesterol). The sterol content of the common aquaculture feed alga *Pavlova lutheri* is as follows: 20 µg/g β-sitosterol, 13 µg/g stigmasterol, and 6 µg/g campesterol. *Nannochloropsis oceanica* contained 36.6, 13.54 and 3.16 µg/g of ergosterol, fucosterol and campesterol, respectively. *I. galbana* consisted of 14.46 µg/g of fucosterol, and *P. pinguis* attained a total sterol content of 17% with 5% stigmasterol (Fernandes et al., 2020; Hikihara et al., 2020). Incorporation of this sterol-rich alga in the diets increased the sterol contents of six bivalve molluscs in aquaculture (Hikihara et al., 2020).

5.3.3 Pigments and Antioxidants

The critical microalgae-derived pigment for aquaculture feed is astaxanthin (AX) derived from *Haematococcus pluvialis*. AX is used as a colorant for bright pink color development in salmonids, shrimp, crabs, and ornamental fish. It stimulates an immune response in the farmed fish protecting against infections and elicits antioxidant and anti-inflammatory properties (Bjerkeng et al., 2007). Atlantic salmon is specifically sensitive to AX availability (Ytrestøyl et al., 2021). Diets low in AX led to mild inflammation, with marked differences in gene expression. A total of 553 genes were differentially expressed in the liver, skeletal muscle, and intestines, of which 119 related to stress response and inflammation were upregulated. Significant changes in intestinal functions were observed, including malabsorption of lipids and downregulation of immune-related genes, while the fat content of the liver was elevated (Ytrestøyl et al., 2021). In addition, synthetic AX is functionally inferior to AX derived from *H. pluvialis*. AX from *H. pluvialis* showed 50-fold and 20-fold higher singlet oxygen quenching and free radical scavenging activity, respectively. The occurrence of a single enantiomeric esterified form (3S-3′S) of AX, along with other alga-derived carotenoids, such as β-carotene, lutein, and canthaxanthin, were attributed to the superior activity (Capelli et al., 2013). The Chinese mitten crab (*Eriocheir sinensis*) preferentially accumulated the 3S-3′S enantiomer of AX in the ovaries, and synthetic AX was required in significantly higher doses to attain a similar effect (Su et al., 2020). AX has additional immune-boosting activities and protects against common aquaculture pathogens such as *Vibrio* sp. (Lim et al., 2021). The presence of other carotenoids in microalgal biomass is responsible for the improved response of aquaculture species against oxidative stress due to the antioxidant potential.

Other lesser-known compounds that could confer antioxidant activity are the polyphenolic compounds, such as phenolic acids, lignans, flavonoids, flavones, isoflavones, flavonols, stilbenes, and anthocyanins present in microalgae. The common green alga *C. vulgaris* consisted of polyphenols such as gallic acid, caffeic acid, p-coumaric acid, and ferulic acid at a concentration of 0.13–0.86 mg/g (Wan Mahmood et al., 2019). Biomass of *Fischerella ambigua, Dunaliella salina, Oocystis pusilla, Nostoc muscorum*, and *Scenedesmus rubescent* was shown to contain 9.62–48.57 mg/g gallic acid, which is responsible for the higher antioxidant activity demonstrated by these algae (Morowvat & Ghasemi, 2016). Regular aquaculture feed diatoms such as *Phaeodactylum, Isochrysis*, and *Tetraselmis* biomass consisted of >3 mg/g gallic acid, which conferred higher antioxidant activity (Goiris et al., 2012). *Scenedesmus* sp. with enhanced antioxidant activity was shown to contain polyphenols such as rutin and quercetin at 0.11 and 0.85 mg/g, respectively (Bulut et al., 2019). A rare phenolic compound named apigenin (4′, 5, 7,-trihydroxyflavone) isolated from a Moroccan strain of *Arthrospira platensis* showed potent radical scavenging activity (Bellahcen et al., 2020).

In conclusion, microalgal biomass is rich in dietary fiber and other functional components to be used as a filler feed. Table 5.1 summarizes the use of microalgal meal in aquaculture feed, and the benefits range from improvement in growth parameters to marked resistance against aquatic pathogens.

TABLE 5.1
Summary of the Use of Various Microalgae as Feed in Aquaculture

Microalgal Source	Aquatic Species	Treatment Method and Dosage	Beneficial Effects	Reference
Nannochloropsis gaditana—crude extract	Senegalese sole—*Solea senegalensi*	Larvae treated with algal extract for 2 h and cultivated for 32 days post-hatch, followed by immune challenge with lymphocystis disease virus (LCDV)	—Increase in expression levels of chemokines and antiviral transcripts, heightened immune response against viral infection —Reduction in pro-inflammatory cytokines —Suitable for larval programming of antiviral immunity	(Carballo et al., 2020)

(Continued)

TABLE 5.1 (CONTINUED)

Microalgal Source	Aquatic Species	Treatment Method and Dosage	Beneficial Effects	Reference
Arthrospira platensis	Coral trout— *Plectropomus leopardus*	0–10% algal meal in fish feed for 8 weeks, followed by *Vibrio harveyi* challenge	—Significant improvement in specific growth rate, weight gain —Improved hematological profile —Improved stress response —Improved immune response —Increased lysozyme activity, respiratory bursts, and immunoglobulin levels —Survival rate against *Vibrio harveyi* infection increased from 40% in the control group to 80% in the 10% algal meal supplemented group	(Yu et al., 2018)
Oedocladium sp. and *Tribonema* sp.	Gibel carp— *Carassius auratus gibelio*	40 g/kg *Oedocladium* sp. meal or 50 g/kg *Tribonema* sp. meal in a basal diet for 40 days	—No difference in growth parameters or feed efficiency —Improved yellowness of the fillet —Higher PUFA content compared to control, especially DHA, DPA, and the ratio of n-3/n-6 PUFAs	(Chen et al., 2019)
Defatted protein-rich *Nannochloropsis oculata* and DHA-rich *Schizochytrium* sp.	Nile tilapia— *Oreochromis niloticus*	3–8% of *Nannochloropsis* and 3.2% of *Schizochytrium* sp. in feed for 184 days	—Significant improvement in growth rate and weight gain compared to the control group ($p < 0.05$) —1.8% higher lipid content in fillets —Higher DHA deposition in fish (5.15 mg/g fillet) in microalgae-fed fish compared to control (2.47 mg/g fillet) —The combination of defatted protein meal and algal oil can replace fishmeal and fish oil and is economically feasible	(Sarker et al., 2020)
Defatted biomass of *Desmodesmus* sp.	Atlantic salmon—*Salmo salar*	10–20% by weight algal meal in fish feed for 70 days	—No significant detrimental effects or decrease in growth rate upon replacement of fishmeal with an algal meal —Minor difference in distal intestinal proteome —High feasibility of fishmeal replacement in Atlantic salmon feed	(Kiron et al., 2016)

TABLE 5.1 (CONTINUED)

Microalgal Source	Aquatic Species	Treatment Method and Dosage	Beneficial Effects	Reference
Dunaliella extract (Algro Natural)	Black tiger shrimp—*Penaeus monodon*	125–300 mg/Kg in basal feed for 8 weeks	—100% survival rates and improved weight gain —Enhanced color intensity of boiled shrimp from orange to orange-red —Enhancement of total carotenoid and astaxanthin content in mature shrimp —Improved resistance against WSSV —Improved stress tolerance against low dissolved oxygen conditions	(Supamattaya et al., 2005)

5.4 MARINE MACROALGAE AS A POTENTIAL FILLER FEED INGREDIENT

Macroalgae are photosynthetic, multicellular plant-like organisms prevalently seen in marine habitats, hence the term "seaweeds." Broadly, macroalgae can be classified into three types based on the pigmentation type and color: *Chlorophyta* or green macroalgae, colored by chlorophylls; *Rodophyta* or red macroalgae, colored by phycoerythrin or phycocyanin; and finally *Phaeophyta* or brown macroalgae, colored by fucoxanthin. Annual global macroalgal biomass production is at 34.7 million tonnes in 2019, of which green macroalgae account for only 1%, while red and brown macroalgae account for the majority due to the varied applications of their polysaccharides as hydrocolloids in various industries (Ferdouse et al., 2018). Macroalgal biomass typically contains carbohydrates, proteins, lipids, moisture, and ash, in addition to high quantities of minerals. Certain aquatic animals, such as abalone, solely rely on macroalgal diets for farming (Daume, 2009). Major health benefits include immune modulation, enhanced growth parameters, and prebiotic effect (Table 5.2). While microalgal meals have been tried to replace fishmeal and fish oils, macroalgal meals have been used as supplements in lower quantities for positive effects on farmed animals. This is due to the presence of several unique, non-digestible polysaccharides in macroalgae that act as immune-modulating agents. Due to their inert nature, they can serve well as a filler feed as well.

Macroalgal polysaccharides account for over 30–75% of biomass composition by weight and are by far the most explored component of the biomass. The carbohydrates can be structural polysaccharides present in cell walls or storage polysaccharides present in the cells. All macroalgae contain specific cell wall polymers, such as cellulose, hemicellulose, and lignin in varying compositions. Starch or floridean starch is frequently the storage polysaccharide (Jönsson et al., 2020). Nevertheless, each class of macroalgae has specific structural and functional polysaccharides, which contribute to the diversity of macroalgal polysaccharides. Green macroalgae, with *Ulva* or sea lettuce as the representative species, possess ulvan as a heterogenous polysaccharide in their cell walls. Ulvans are composed of rhamnose and sugar organic acids such as guluronic acid and iduronic acids (Steinbruch et al., 2020). Red macroalgae are highly cultivated macroalgae for the extraction of hydrocolloids such as agar and carrageenan. The major monosaccharides present in red macroalgal hydrolysates are galactose and anhydro galactose, of which anhydro galactose is non-metabolizable (Nallasivam et al., 2022). Brown macroalgae contain various polysaccharides such as alginates, laminarin, and fucoidan in their cell walls. The monosaccharides seen in brown algal polysaccharides include fucose, mannitol, glucose, mannuronic acid, and glucuronic acid (Li et al., 2019; Yin & Wang, 2018). The unique feature of macroalgal polysaccharides that makes them a potential feed ingredient is their immune-stimulating properties, as listed in Table 5.3. Such immune

TABLE 5.2
Summary of the Use of Macroalgal Biomass as a Feed Ingredient in Aquaculture

Macroalgal biomass	Aquatic species	Dosage and Treatment Method	Beneficial Effects	Reference
Red seaweed—*Asparagopsis taxiformis* powder and methanolic extract	Atlantic salmon—*Salmo salar*	3% biomass powder and/or 0.6–1.2% methanolic extract fed for 4 weeks	—8% increase in feed intake —26% increase in weight gain —Increment in probiotic *Shewanella* sp. (ASV16) in fish gut —Reduction in serum glucose level —Increase in innate immune response	(Thépot et al., 2022)
Sargassum polycystum	Asian sea bass—*Lates calcarifer*	1–4.5% by weight in feed for 55 days	—Potent prebiotic; increased the levels of *Lactobacillus paracasei* probiotic in intestines of seaweed-supplemented fish —Improved survival rate, growth rate, and feed conversion efficiency —Increased protein and iron content in carcass meat	(Nazarudin et al., 2020)
Gracilariopsis persica, Hypnea flagelliformis, Sargassum boveanum	Rainbow trout—*Oncorhynchus mykiss*	5–10% by weight of individual seaweed powder in feed for 83 days	—Improved serum immunity indicators —Enhanced oxidative stress response —Stimulation of immune-related gene expression, including various cytokines such as interleukins, tumor necrosis factor, and lysozyme	(Vazirzadeh et al., 2020)
Extract of the seaweed mixture—*Ulva lactuca, Jania rubens*, and *Pterocladia capillacea*	Striped catfish—*Pangasianodon hypophthalmus*	0–3% seaweed extract in feed for 60 days and challenged with the pathogen *Aeromonas hydrophila*	—Improved survival rate after pathogen challenge by 66% —Significant improvement in the immune response: increased serum lysozyme, immunoglobulin, and complement levels —Improvement in feed intake and specific growth rate —Improvement in hematological parameters and biochemical markers indicating general good health —Potent antioxidant response—indicated by reduced lipid peroxidation	(Abdelhamid et al., 2021)

TABLE 5.2 (CONTINUED)

Macroalgal biomass	Aquatic species	Dosage and Treatment Method	Beneficial Effects	Reference
Porphyra haitanensis, Undaria pinnatifida, Saccharina japonica, Gracilaria lemaneiformis	Pacific white shrimp—*Litopenaeus vannamei*	2% by weight in the basal diet for 8 weeks followed by challenge with white spot syndrome virus (WSSV)	—Improved specific growth rate and protein conversion ratio —Improved oxidative stress response —Increase in beneficial gut bacteria *Bacteroidetes, Firmicutes*, and *Bacillaceae* and decrease in harmful gut bacteria *Gammaproteobacteria* and *Vibrionaceae* —Protection of liver and pancreas —Significant reduction in mortality followed by WSSV challenge	(Niu et al., 2018)
Polysaccharides derived from *Enteromorpha prolifera*	Banana shrimp—*Fenneropenaeus merguiensis*	0–3 g/Kg polysaccharides in a basal diet for 42 days	—Improved weight gain —Improved antioxidant activity and decrease in lipid peroxidation —Changes in intestinal gut bacteria—increased levels of *Firmicutes* and decreased levels of *Vibrio* sp. —Improved expression of immune-related proteins, including lectin, antioxidant enzymes, transcription factors, and lysozymes in hepatopancreas, gill, and intestines	(Liu et al., 2020)
Red seaweed *Kappaphycus alvarezii*	Pacific white shrimp—*Litopenaeus vannamei*	15 g/Kg seaweed powder in the basal diet for 30 days followed by *Vibrio* challenge	—Improved weight gain —More than 10% improvement in survival rate followed by *Vibrio* challenge —Significant reduction in cell lesions followed by *Vibrio* infection	(Suantika et al., 2018)
Solvent extracts of red seaweed *Gracilaria fisheri*	Black tiger shrimp—*Penaeus monodon*	Injection of 0.5–1 mg/Kg ethanolic extract of red seaweed twice, followed by regular diet for 14 days and challenged with *Vibrio*	—Cumulative mortality from *Vibrio* challenge was reduced from 100% in control to 17.5% in 1 mg/Kg red seaweed extract injection —Similar effect was observed in juveniles —Marked increase in antimicrobial activity and immune response compared to control	(Kanjana et al., 2011)

TABLE 5.3
Major Macroalgal Polysaccharides and Their Specific Health Benefits in Aquaculture Feed

Macroalgae Derived Polysaccharide	Source	Structural Aspects*	Positive Health Effects in Aquatic Animals	References
Ulvan	Green macroalgae of the genus *Ulva*	Sulfated heteropolysaccharide is composed mainly of rhamnose, uronic acids such as glucuronic acid and iduronic acid, and xylose	Immunostimulatory compound and a potential vaccine adjuvant, immunity against aquaculture pathogens, improved innate immune response, and reduced mortality rates post-infection	(Declarador et al., 2014)
Fucoidan	Cell walls of brown seaweeds	Sulfated polysaccharides rich in fucose and other monosaccharides including galactose, glucose, mannose, xylose, and uronic acids	Potent immunostimulatory agent is effective in both non-specific and cell-mediated immunity, enhanced growth parameters and overall metabolic health, protection against common aquaculture pathogens, antioxidant activity	(Sony et al., 2019)
Alginate	Cell walls of brown seaweeds	Linear homo- or heteropolymer of $(1\to4)$ linked α-L-guluronic acid (G) and β-D-mannuronic acid (M), the composition is GG or MM or MG	Potent immune stimulation, non-specific innate immune response activation, protection against aquaculture pathogens, prebiotic activity, improved metabolic health, and improved growth parameters. Binders and encapsulation agents in feed formulation	(Ashouri et al., 2020)
Laminarin	Storage polysaccharides in brown seaweeds	A $\beta(1\to3)$ glucan with $\beta(1\to6)$ branches is composed mainly of glucose units	Immunostimulatory compound activating cell-mediated response and cytokine expression, enhanced growth parameters and reduced feed coefficient, improved feed efficiency, reduces energy decomposition, and prebiotic activity	(Jiang et al., 2021)
Agar	Cell walls of red seaweeds	Composed of agarose and agaropectin. $(1\to4)$ linked β-D-galactose and 3,6-anhydro-α-L-galactose, with sulfate, pyruvate, and methyl groups. Agarose is linear, while agaropectin is branched	Binding agent in feed formulation. Aids in synchronization of reproductive cycles of sea urchins (*Paracentrotus lividus*) in intensive farming conditions. Improvement in growth parameters and immune response of farmed fish	(Fabbrocini, 2015)

Filler Feed from Marine Micro- and Macroalgae

TABLE 5.3 (CONTINUED)

Macroalgae Derived Polysaccharide	Source	Structural Aspects*	Positive Health Effects in Aquatic Animals	References
Carrageenan	Cell walls of red macroalgae	Linear polymer composed of alternating units of β-D-galactose and 3,6-anhydro-α-D-galactose linked by α-(1→3) and β-(1→4) bonds	Potential immune stimulatory effect confers resistance against aquaculture pathogens including viruses, improved growth parameters and survival rate, confers stress tolerance, and prebiotic effect	(Mariot et al., 2021)

* The composition given in this table is a general overview, and the specific composition of these polysaccharides is subject to variations based on the harvest season, macroalgal source, and habitat of the source.

modulation confers resistance against common aquaculture pathogens such as *Vibrio harveyii, Aeromonas hydrophila*, and the white spot syndrome virus. It also boosts the survival rate of reared fish with improved disease resistance.

Rohu fish *Labeo rohita* juveniles fed with fucoidan-rich seaweeds and low doses of L-methionine (2% algal meal and 0.9% methionine) showed a significant improvement in feed conversion ratio, weight gain, specific growth rate, and protein efficiency ratio compared to a control group. Challenge with the pathogenic *Aeromonas hydrophila* showed a higher survival rate of 70% compared to 30% in the control group (Mir et al., 2017). Nile tilapia juveniles fed with purified kappa-carrageenan from *Hypnea musciformis* (up to 10 g/Kg for 30 days) showed an increased survival rate from 88% in the control group to 96% in the carrageenan group. Carrageenan-fed fish showed reduced mortality by 25% compared to the control group when challenged with the aquaculture pathogen *Edwardsiella tarda* (Villamil et al., 2019). The black tiger shrimp *Peneaus monodon* showed prolonged survival after white spot syndrome virus challenge when fed with diets containing 500–1500 mg/kg ulvan extract. Total hemocyte count, respiratory burst activity, and phenoloxidase activity are significantly higher in shrimp fed ulvan (Declarador et al., 2014).

Prebiotics are non-digestible food components that are selectively fermented in the intestine to bring about appropriate changes in the gut microbial population and improve the overall health and well-being of the animal. Prebiotics act as stimulants and increase the relative abundance of probiotic bacteria such as *Lactobacillus, Bifidobacterium, Enterococcus, Bacillus, Streptococcus*, and *Escherichia* while reducing the number of *Clostridium, Proteus, Pseudomonas, Salmonella*, and *Staphylococcus*. Prebiotics also prevent the effective colonization of the intestines by potentially pathogenic bacteria and are capable of bringing about structural and physiological changes in the intestines via differential gene expression (Roberfroid, 2008). Furthermore, prebiotics can increase the growth rate of aquatic animals and reduce the feed conversion ratio, thereby decreasing the amount of feed required for the production of 1 kg of fillet (Ganguly et al., 2013). Prebiotics are especially essential in animals raised via intense farming practices, which impose severe environmental stress and poses a higher infection risk. Macroalgal polysaccharides can function as prebiotics and improve the intestinal health of aquaculture animals. *Grateloupia and Eucheuma* polysaccharides enhanced the growth of bifidobacteria at all tested concentrations from 0.1–0.5% (Chen et al., 2018). A 0.5% inclusion of the alginate-derived oligosaccharides increased the abundance of short-chain fatty acid-producing *Aliivibrio logei, Aquabacterium parvum, Bacillus andersonii*, and *Achromobacter insolitus* in the intestines of commercially important Atlantic salmon (Gupta et al., 2019). Thus, seaweed polysaccharides can be applied as potential prebiotic agents in aquaculture feed.

5.5 CONCLUSIONS AND PERSPECTIVES

Marine algal biomass has appropriate biomass composition and valuable techno-functional components to add high value as an aquaculture feed ingredient. Microalgae are traditionally used as aquaculture feeds, particularly in larval hatcheries where specific larvae feed exclusively on microalgae. Microalgae-derived pigment astaxanthin is essential for the growth, immunity, and metabolic health of commercially important fish such as salmonids. On the other hand, microalgal biomass is energetically dense with bioavailable carbohydrates and proteins, and its application as a filler feed ingredient might interfere with the net calorific value and nutritional content of the feed. These factors need to be taken into consideration while formulating the feed. Spent microalgal biomass such as lipid-extracted or pigment-extracted residue is a potential source for filler feed. Proximate composition analysis and knowledge regarding the source of the biomass are crucial before incorporating them as filler feed. Conversely, macroalgae contain high amounts of non-digestible polysaccharides, which are more suitable as a dietary fiber and filler feed ingredient. The processing of red and brown macroalgal biomass for hydrocolloid extraction leaves behind a carbohydrate-rich residue, which could be used as a filler feed. In both microalgae and macroalgae, the processed biomass residue will contain peptides, oligosaccharides, pigments, and fatty acids, which will provide additional health benefits. For now, algal biomass is used in aquaculture feed as an expensive added-value feed supplement for specific reasons such as fish oil replacement or enhanced coloration. Economic cultivation of algal biomass, with advanced technologies and higher productivity, will enable its cost-competitive use in aquaculture feeds as filler feeds or prebiotics.

REFERENCES

Abdelhamid, A.F., Ayoub, H.F., Abd El-Gawad, E.A., Abdelghany, M.F., Abdel-Tawwab, M. 2021. Potential effects of dietary seaweeds mixture on the growth performance, antioxidant status, immunity response, and resistance of striped catfish (*Pangasianodon hypophthalmus*) against *Aeromonas hydrophila* infection. *Fish & Shellfish Immunology*, **119**, 76–83.

Ashouri, G., Mahboobi Soofiani, N., Hoseinifar, S.H., Jalali, S.A.H., Morshedi, V., Valinassab, T., Bagheri, D., Van Doan, H., Torfi Mozanzadeh, M., Carnevali, O. 2020. Influence of dietary sodium alginate and *Pediococcus acidilactici* on liver antioxidant status, intestinal lysozyme gene expression, histomorphology, microbiota, and digestive enzymes activity, in Asian sea bass (*Lates calcarifer*) juveniles. *Aquaculture*, **518**, 734638.

Barreto-Curiel, F., Focken, U., D'Abramo, L.R., Mata-Sotres, J., Viana, M.T. 2019. Assessment of amino acid requirements for *Totoaba macdonaldi* at different levels of protein using stable isotopes and a non-digestible protein source as a filler. *Aquaculture*, **503**, 550–561.

Bellahcen, T.O., Aamiri, A., Touam, I., Hmimid, F., Amrani, A.E., Cherif, A., Cherki, M. 2020. Evaluation of Moroccan microalgae: *Spirulina platensis* as a potential source of natural antioxidants. *Journal of Complementary and Integrative Medicine*, **17**(3).

Bjerkeng, B., Peisker, M., von Schwartzenberg, K., Ytrestøyl, T., Åsgård, T. 2007. Digestibility and muscle retention of astaxanthin in Atlantic salmon, *Salmo salar*, fed diets with the red yeast *Phaffia rhodozyma* in comparison with synthetic formulated astaxanthin. *Aquaculture*, **269**(1), 476–489.

Boyd, C.E., McNevin, A.A., Davis, R.P. 2022. The contribution of fisheries and aquaculture to the global protein supply. *Food Security*, **14**, 805–827. https://doi.org/10.1007/s12571-021-01246-9.

Bromley, P.J., Adkins, T.C. 1984. The influence of cellulose filler on feeding, growth and utilization of protein and energy in rainbow trout, *Salmo gairdnerii* Richardson. *Journal of Fish Biology*, **24**(2), 235–244.

Bulut, O., Akın, D., Sönmez, Ç., Öktem, A., Yücel, M., Öktem, H.A. 2019. Phenolic compounds, carotenoids, and antioxidant capacities of a thermo-tolerant *Scenedesmus* sp. (Chlorophyta) extracted with different solvents. *Journal of Applied Phycology*, **31**(3), 1675–1683.

Capelli, B., Bagchi, D., Cysewski, G.R. 2013. Synthetic astaxanthin is significantly inferior to algal-based astaxanthin as an antioxidant and may not be suitable as a human nutraceutical supplement. *Nutrafoods*, **12**(4), 145–152.

Carballo, C., Mateus, A.P., Maya, C., Mantecón, L., Power, D.M., Manchado, M. 2020. Microalgal extracts induce larval programming and modify growth and the immune response to bioactive treatments and LCDV in Senegalese sole post-larvae. *Fish & Shellfish Immunology*, **106**, 263–272.

Carvalho, M., Montero, D., Rosenlund, G., Fontanillas, R., Ginés, R., Izquierdo, M. 2020. Effective complete replacement of fish oil by combining poultry and microalgae oils in practical diets for gilthead sea bream (*Sparus aurata*) fingerlings. *Aquaculture*, **529**, 735696.

Chen, W., Wang, Y., Han, D., Zhu, X., Xie, S., Han, D., Hu, Q. 2019. Two filamentous microalgae as feed ingredients improved flesh quality and enhanced antioxidant capacity and immunity of the gibel carp (*Carassius auratus gibelio*). *Aquaculture Nutrition*, **25**(5), 1145–1155.

Chen, X., Sun, Y., Hu, L., Liu, S., Yu, H., Xing, R.E., Li, R., Wang, X., Li, P. 2018. In vitro prebiotic effects of seaweed polysaccharides. *Journal of Oceanology and Limnology*, **36**(3), 926–932.

Daume, S. 2009. The roles of bacteria and micro and macro algae in abalone aquaculture: A review. *Journal of Shellfish Research*, **25**, 151–157.

Declarador, R., Serrano Jr, A., Corre, V. 2014. Ulvan extract acts as immunostimulant against white spot syndrome virus (WSSV) in juvenile black tiger shrimp *Penaeus monodon*. *AACL Bioflux*, **7**, 153–161.

Dias, J., Huelvan, C., Dinis, M.T., Métailler, R. 1998. Influence of dietary bulk agents (silica, cellulose and a natural zeolite) on protein digestibility, growth, feed intake and feed transit time in European seabass (*Dicentrarchus labrax*) juveniles. *Aquatic Living Resources*, **11**(4), 219–226.

Durazo, E., Cruz, A.C., López, L.M., Lazo, J.P., Drawbridge, M., Viana, M.T. 2010. Effects of digestible protein levels in isonitrogenous diets on growth performance and tissue composition of juvenile *Atractoscion nobilis*. *Aquaculture Nutrition*, **16**(1), 54–60.

Fabbrocini, A. 2015. Agar-based biocomposites slow down progression in the reproductive cycle facilitating synchronization of the gonads of reared specimens of *Paracentrotus lividus*. *International Journal of Aquaculture and Fishery Sciences*, **1**, 35–41.

Ferdouse, F., Holdt, S.L., Smith, R., Murúa, P., Yang, Z. 2018. The global status of seaweed production, trade and reutilization. *FAO Globefish Research Programme*, **124**. http://www.fao.org/3/CA1121EN/ca1121en.pdf.

Fernandes, T., Martel, A., Cordeiro, N. 2020. Exploring *Pavlova pinguis* chemical diversity: A potentially novel source of high value compounds. *Scientific Reports*, **10**(1).

Food and Agriculture Organization of the United Nations. 2020. *The State of World Fisheries and Aquaculture*. https://doi.org/10.4060/ca9231en.

Ganguly, S., Dora, K.C., Sarkar, S., Chowdhury, S. 2013. Supplementation of prebiotics in fish feed: A review. *Reviews in Fish Biology and Fisheries*, **23**(2), 195–199.

German-Báez, L.J., Valdez-Flores, M.A., Félix-Medina, J.V., Norzagaray-Valenzuela, C.D., Santos-Ballardo, D.U., Reyes-Moreno, C., Shelton, L.M., Valdez-Ortiz, A. 2017. Chemical composition and physicochemical properties of *Phaeodactylum tricornutum* microalgal residual biomass. *Food Science and Technology International*, **23**(8), 681–689.

Goiris, K., Muylaert, K., Fraeye, I., Foubert, I., De Brabanter, J., De Cooman, L. 2012. Antioxidant potential of microalgae in relation to their phenolic and carotenoid content. *Journal of Applied Phycology*, **24**(6), 1477–1486.

Gong, Y., Sørensen, S.L., Dahle, D., Nadanasabesan, N., Dias, J., Valente, L.M.P., Sørensen, M., Kiron, V. 2020. Approaches to improve utilization of *Nannochloropsis oceanica* in plant-based feeds for Atlantic salmon. *Aquaculture*, **522**, 735122.

Gupta, S., Lokesh, J., Abdelhafiz, Y., Siriyappagouder, P., Pierre, R., Sørensen, M., Fernandes, J.M.O., Kiron, V. 2019. Macroalga-derived alginate oligosaccharide alters intestinal bacteria of Atlantic salmon. *Frontiers in Microbiology*, **10**.

Hart, B., Schurr, R., Narendranath, N., Kuehnle, A., Colombo, S.M. 2021. Digestibility of *Schizochytrium* sp. whole cell biomass by Atlantic salmon (*Salmo salar*). *Aquaculture*, **533**, 736156.

Hikihara, R., Yamasaki, Y., Shikata, T., Nakayama, N., Sakamoto, S., Kato, S., Hatate, H., Tanaka, R. 2020. Analysis of phytosterol, fatty acid, and carotenoid composition of 19 microalgae and 6 bivalve species. *Journal of Aquatic Food Product Technology*, **29**(5), 461–479.

Hoseinifar, S.H., Fazelan, Z., Bayani, M., Yousefi, M., Van Doan, H., Yazici, M. 2022. Dietary red macroalgae (*Halopithys incurva*) improved systemic an mucosal immune and antioxidant parameters and modulated related gene expression in zebrafish (Danio rerio). *Fish & Shellfish Immunology*, **123**, 164–171.

Jiang, H., Wang, M., Zheng, Y., Chen, F., Fu, L., Zhong, L., Chen, X., Bian, W. 2021. Dietary laminarin administration to enhance the immune responses, promote growing and strengthen physique in *Ictalurus punctatus*. *Aquaculture Nutrition*, **27**(4), 1181–1191.

Jones, S.W., Karpol, A., Friedman, S., Maru, B.T., Tracy, B.P. 2020. Recent advances in single cell protein use as a feed ingredient in aquaculture. *Current Opinion in Biotechnology*, **61**, 189–197.

Jönsson, M., Allahgholi, L., Sardari, R.R.R., Hreggviðsson, G.O., Nordberg Karlsson, E. 2020. Extraction and modification of macroalgal polysaccharides for current and next-generation applications. *Molecules*, **25**(4).

Kanjana, K., Radtanatip, T., Asuvapongpatana, S., Withyachumnarnkul, B., Wongprasert, K. 2011. Solvent extracts of the red seaweed *Gracilaria fisheri* prevent *Vibrio harveyi* infections in the black tiger shrimp Penaeus monodon. *Fish & Shellfish Immunology*, **30**(1), 389–396.

Kapaun, E., Reisser, W. 1995. A chitin-like glycan in the cell wall of a *Chlorella* sp. (Chlorococcales, Chlorophyceae). *Planta*, **197**(4), 577–582.

Kiron, V., Sørensen, M., Huntley, M., Vasanth, G.K., Gong, Y., Dahle, D., Palihawadana, A.M. 2016. Defatted biomass of the microalga, *Desmodesmus* sp., can replace fishmeal in the feeds for Atlantic salmon. *Frontiers in Marine Science*, **3**.

Kusmayadi, A., Leong, Y.K., Yen, H.-W., Huang, C.-Y., Chang, J.-S. 2021. Microalgae as sustainable food and feed sources for animals and humans—biotechnological and environmental aspects. *Chemosphere*, **271**, 129800.

Li, S.-Y., Wang, Z.-P., Wang, L.-N., Peng, J.-X., Wang, Y.-N., Han, Y.-T., Zhao, S.-F. 2019. Combined enzymatic hydrolysis and selective fermentation for green production of alginate oligosaccharides from *Laminaria japonica*. *Bioresource Technology*, **281**, 84–89.

Lim, K.C., Yusoff, F.M., Shariff, M., Kamarudin, M.S. 2021. Dietary astaxanthin augments disease resistance of Asian seabass, *Lates calcarifer* (Bloch, 1790), against *Vibrio alginolyticus* infection. *Fish & Shellfish Immunology*, **114**, 90–101.

Liu, W.-C., Zhou, S.-H., Balasubramanian, B., Zeng, F.-Y., Sun, C.-B., Pang, H.-Y. 2020. Dietary seaweed (*Enteromorpha*) polysaccharides improves growth performance involved in regulation of immune responses, intestinal morphology and microbial community in banana shrimp *Fenneropenaeus merguiensis*. *Fish & Shellfish Immunology*, **104**, 202–212.

Mariot, L.V., Bolívar, N., Coelho, J.D.R., Goncalves, P., Colombo, S.M., do Nascimento, F.V., Schleder, D.D., Hayashi, L. 2021. Diets supplemented with carrageenan increase the resistance of the Pacific white shrimp to WSSV without changing its growth performance parameters. *Aquaculture*, **545**, 737172.

Mir, I.N., Sahu, N.P., Pal, A.K., Makesh, M. 2017. Synergistic effect of l-methionine and fucoidan rich extract in eliciting growth and non-specific immune response of *Labeo rohita* fingerlings against *Aeromonas hydrophila*. *Aquaculture*, **479**, 396–403.

Morowvat, M.H., Ghasemi, Y. 2016. Evaluation of antioxidant properties of some naturally isolated microalgae: Identification and characterization of the most efficient strain. *Biocatalysis and Agricultural Biotechnology*, **8**, 263–269.

Müller, M., Canfora, E.E., Blaak, E.E. 2018. Gastrointestinal transit time, glucose homeostasis and metabolic health: Modulation by dietary fibers. *Nutrients*, **10**(3).

Nagarajan, D., Varjani, S., Lee, D.-J., Chang, J.-S. 2021. Sustainable aquaculture and animal feed from microalgae—nutritive value and techno-functional components. *Renewable and Sustainable Energy Reviews*, **150**, 111549.

Nallasivam, J., Francis Prashanth, P., Harisankar, S., Nori, S., Suryanarayan, S., Chakravarthy, S.R., Vinu, R. 2022. Valorization of red macroalgae biomass via hydrothermal liquefaction using homogeneous catalysts. *Bioresource Technology*, **346**, 126515.

Nazarudin, M.F., Yusoff, F., Idrus, E.S., Aliyu-Paiko, M. 2020. Brown seaweed *Sargassum polycystum* as dietary supplement exhibits prebiotic potentials in Asian sea bass *Lates calcarifer* fingerlings. *Aquaculture Reports*, **18**, 100488.

Niccolai, A., Chini Zittelli, G., Rodolfi, L., Biondi, N., Tredici, M.R. 2019. Microalgae of interest as food source: Biochemical composition and digestibility. *Algal Research*, **42**, 101617.

Niu, J., Xie, J.J., Guo, T.Y., Fang, H.H., Zhang, Y.M., Liao, S.Y., Xie, S.W., Liu, Y.J., Tian, L.X. 2018. Comparison and evaluation of four species of macro-algaes as dietary ingredients in *Litopenaeus vannamei* under normal rearing and WSSV challenge conditions: Effect on growth, immune response, and intestinal microbiota. *Frontiers in Physiology*, **9**, 1880.

Parrish, C.C. 2009. Essential fatty acids in aquatic food webs. in: *Lipids in Aquatic Ecosystems*, (Eds.) M. Kainz, M.T. Brett, M.T. Arts, Springer, New York, NY, pp. 309–326.

Pina-Pérez, M.C., Brück, W.M., Brück, T., Beyrer, M. 2019. Chapter 4—microalgae as healthy ingredients for functional foods. in: *The Role of Alternative and Innovative Food Ingredients and Products in Consumer Wellness*, (Ed.) C.M. Galanakis, Academic Press, pp. 103–137.

Ranglová, K., Bureš, M., Manoel, J.C., Lakatos, G.E., Masojídek, J. 2022. Efficient microalgae feed production for fish hatcheries using an annular column photobioreactor characterized by a short light path and central LED illumination. *Journal of Applied Phycology*, **34**(1), 31–41.

Roberfroid, M.B. 2008. Chapter 1—general introduction: Prebiotics in nutrition. in: *Handbook of Prebiotics* (1st ed.), (Eds.) G.R. Gibson, M. Roberfroid, CRC Press, pp. 1–12. https://doi.org/10.1201/9780849381829.

Sadovskaya, I., Souissi, A., Souissi, S., Grard, T., Lencel, P., Greene, C.M., Duin, S., Dmitrenok, P.S., Chizhov, A.O., Shashkov, A.S., Usov, A.I. 2014. Chemical structure and biological activity of a highly branched (1→3, 1→6)-β-d-glucan from *Isochrysis galbana*. *Carbohydrate Polymers*, **111**, 139–148.

Sarker, P.K., Kapuscinski, A.R., McKuin, B., Fitzgerald, D.S., Nash, H.M., Greenwood, C. 2020. Microalgae-blend tilapia feed eliminates fishmeal and fish oil, improves growth, and is cost viable. *Scientific Reports*, **10**(1), 19328.

Satessa, G.D., Kjeldsen, N.J., Mansouryar, M., Hansen, H.H., Bache, J.K., Nielsen, M.O. 2020. Effects of alternative feed additives to medicinal zinc oxide on productivity, diarrhoea incidence and gut development in weaned piglets. *Animal*, **14**(8), 1638–1646.

Sony, N.M., Ishikawa, M., Hossain, M.S., Koshio, S., Yokoyama, S. 2019. The effect of dietary fucoidan on growth, immune functions, blood characteristics and oxidative stress resistance of juvenile red sea bream, *Pagrus major*. *Fish Physiology and Biochemistry*, **45**(1), 439–454.

Steinbruch, E., Drabik, D., Epstein, M., Ghosh, S., Prabhu, M.S., Gozin, M., Kribus, A., Golberg, A. 2020. Hydrothermal processing of a green seaweed *Ulva* sp. for the production of monosaccharides, polyhydroxyalkanoates, and hydrochar. *Bioresource Technology*, **318**, 124263.

Su, F., Yu, W., Liu, J. 2020. Comparison of effect of dietary supplementation with *Haematococcus pluvialis* powder and synthetic astaxanthin on carotenoid composition, concentration, esterification degree and astaxanthin isomers in ovaries, hepatopancreas, carapace, epithelium of adult female Chinese mitten crab (*Eriocheir sinensis*). *Aquaculture*, **523**, 735146.

Suantika, G., Situmorang, M.L., Khakim, A., Wibowo, I., Aditiawati, P., Suryanarayan, S., Nori, S., Kumar, S., Putri, F. 2018. Effect of red seaweed *Kappaphycus alvarezii* on growth, survival, and disease resistance of Pacific white shrimp Litopenaeus vannamei against *Vibrio harveyi* in the nursery phase. *Journal of Aquaculture Research & Development*, **9**.

Sun, Y., Wang, H., Guo, G., Pu, Y., Yan, B. 2014. The isolation and antioxidant activity of polysaccharides from the marine microalgae *Isochrysis galbana*. *Carbohydrate Polymers*, **113**, 22–31.

Supamattaya, K., Kiriratnikom, S., Boonyaratpalin, M., Borowitzka, L. 2005. Effect of a Dunaliella extract on growth performance, health condition, immune response and disease resistance in black tiger shrimp (*Penaeus monodon*). *Aquaculture*, **248**(1), 207–216.

Suresh, A.V. 2016. Feed formulation software. in: *Aquafeed Formulation*, (Ed.) S.F. Nates, Academic Press, San Diego, CA, pp. 21–31.

Sushytskyi, L., Lukáč, P., Synytsya, A., Bleha, R., Rajsiglová, L., Capek, P., Pohl, R., Vannucci, L., Čopíková, J., Kaštánek, P. 2020. Immunoactive polysaccharides produced by heterotrophic mutant of green microalga *Parachlorella kessleri* HY1 (Chlorellaceae). *Carbohydrate Polymers*, **246**, 116588.

Thépot, V., Campbell, A.H., Rimmer, M.A., Jelocnik, M., Johnston, C., Evans, B., Paul, N.A. 2022. Dietary inclusion of the red seaweed *Asparagopsis taxiformis* boosts production, stimulates immune response and modulates gut microbiota in Atlantic salmon, *Salmo salar*. *Aquaculture*, **546**, 737286.

Vazirzadeh, A., Marhamati, A., Rabiee, R., Faggio, C. 2020. Immunomodulation, antioxidant enhancement and immune genes up-regulation in rainbow trout (*Oncorhynchus mykiss*) fed on seaweeds included diets. *Fish & Shellfish Immunology*, **106**, 852–858.

Villamil, L., Infante Villamil, S., Rozo, G., Rojas, J. 2019. Effect of dietary administration of kappa carrageenan extracted from Hypnea musciformis on innate immune response, growth, and survival of Nile tilapia (*Oreochromis niloticus*). *Aquaculture International*, **27**(1), 53–62.

Wan Mahmood, W.M.A., Lorwirachsutee, A., Theodoropoulos, C., Gonzalez-Miquel, M. 2019. Polyol-based deep eutectic solvents for extraction of natural polyphenolic antioxidants from *Chlorella vulgaris*. *ACS Sustainable Chemistry and Engineering*, **7**(5), 5018–5026.

Yin, Y., Wang, J. 2018. Pretreatment of macroalgal *Laminaria japonica* by combined microwave-acid method for biohydrogen production. *Bioresource Technology*, **268**, 52–59.

Ytrestøyl, T., Afanasyev, S., Ruyter, B., Hatlen, B., Østbye, T.-K., Krasnov, A. 2021. Transcriptome and functional responses to absence of astaxanthin in Atlantic salmon fed low marine diets. *Comparative Biochemistry and Physiology Part D: Genomics and Proteomics*, **39**, 100841.

Yu, W., Wen, G., Lin, H., Yang, Y., Huang, X., Zhou, C., Zhang, Z., Duan, Y., Huang, Z., Li, T. 2018. Effects of dietary *Spirulina platensis* on growth performance, hematological and serum biochemical parameters, hepatic antioxidant status, immune responses and disease resistance of coral trout *Plectropomus leopardus* (Lacepede, 1802). *Fish & Shellfish Immunology*, **74**, 649–655.

6 Marine Collagen
Valorization of Marine Wastes for Health Care and Biomaterials

Grace Sathyanesan Anisha, Anil Kumar Patel and Reeta Rani Singhania

CONTENTS

6.1 Introduction 77
6.2 Sources and Extraction of Marine Collagen... 78
6.3 Properties of Marine Collagen 81
    6.3.1 Amino Acid Composition.. 81
    6.3.2 Mutable Nature 82
    6.3.3 Tensile Strength 82
    6.3.4 Porosity 83
6.4 Biological Action and Health Benefits of Marine Collagen 83
    6.4.1 Antioxidant Action 83
    6.4.2 Antibacterial Action 83
    6.4.3 Anti-Obesogenic and Anti-Diabetic Action 83
    6.4.4 Food Applications.. 84
    6.4.5 Cosmetic Applications 84
6.5 Biomedical Applications of Marine Collagen 84
    6.5.1 Scaffolds for Tissue Engineering and Regenerative Medicine................ 84
    6.5.2 Wound Healing.. 85
6.6 Summary and Conclusions 85
References... 86

6.1 INTRODUCTION

Collagen is the principal structural protein present in the connective tissues and extracellular matrix of animals that helps in maintaining the structural integrity of these tissues. Collagens form a resourceful biopolymer for pharmaceutical, cosmeceutical, biomedical and tissue engineering applications for humans owing to their biocompatible and biodegradable nature (Liu et al. 2022). Though collagens are extensively available in the mammalian tissues of cows and pigs, the presence of α-gal antigen in these tissues has the probability of eliciting anaphylactic immune responses in at least a minor section of the population (Liu et al. 2022; Steinke et al. 2015). On the contrary, fish and fish-based products trigger immune responses in only <1% of the general population (Salvatore et al. 2020). Moreover, unlike collagens from bovine and porcine tissues, marine collagen has no chance of zoonotic transmission of bovine spongiform encephalopathy and transmissible spongiform encephalopathy (Choi et al. 2015). In this milieu, collagens from marine organisms such as fishes, echinoderms, molluscs and poriferans are gaining acceptance. Marine collagens have better chemical and physical durability compared to collagen from bovine and porcine tissues (Xu et al. 2021; Yamada et al. 2014; Yamamoto et al. 2014).

Advancements in captive and capture fisheries and aquaculture practices have driven up the production and consumption of marine food. Consequently, refuse from the marine food processing

DOI: 10.1201/9781003326946-6

industry comprising the skin, scales, fins, bones and cartilages of fishes and other marine animals has also increased noticeably, thereby posing a menace to waste disposal and management. Moreover, inedible fishes and other marine species that might be accidentally caught also add to the waste generated. The processing of these wastes from the marine food processing industry for the extraction of valuable products such as bioactive collagen not only resolves the problem of waste management but also raises their value. This can contribute to the circular bio-economy and make the marine food industry sustainable and environmentally friendly. Marine biorefinery is a growing field in the purview of the blue bio-economy, where aquatic biomass is converted into novel higher-value products with diverse applications for the benefit of humankind.

Marine collagen with biological actions such as antioxidant, anticoagulant, antimicrobial, anticancer, antihypertensive and immunomodulatory properties is used in cosmeceutical and pharmaceutical preparations (Eser & Gozde 2021). The immense variety of invertebrates and vertebrates inhabiting the marine habitat spanning intertidal and deep sea environments form a rich source of biopolymers with promising health and biomedical applications for humans. Given the rich marine biodiversity, there are different types of collagen depending on their source and the functions that they perform (Fassini et al. 2021).

6.2 SOURCES AND EXTRACTION OF MARINE COLLAGEN

Marine collagen is extracted from different species of fishes, sponges, jellyfishes, molluscs and echinoderms. Filleting, the primary step in fish processing, generates solid wastes in the form of bones, skeletons, heads, tails, viscera, skins and scales. These can be valorized for the production of high-quality proteins such as collagen and gelatin instead of being discarded as wastes. The extraction of biopeptides from several species of marine fishes, including tuna, mackerel, tilapia, pollock, sole, anchovy and others, have been reported (Zamorano-Apodaca et al. 2020). Marine sponges and jellyfishes are the most widely used invertebrates for the extraction of collagen. The marine sponges *Chondrosia reniformis*, *Chondrilla nucula*, *Axinella cannabina* and *Suberites carnosus* and the jellyfish *Rhizostoma pulmo* are examples of invertebrate sources of marine collagen (Salvatore et al. 2020).

Among echinoderms, the sea urchin *Paracentrotus lividus*, the starfish *Echinaster sepositus* and the sea cucumbers *Holothuria tubulosa* and *Apostichopus japonicus* (Table 6.1) are used for the extraction of collagen since they have unique mutable collagenous connective tissues (Ferrario et al. 2017). The gonads of the common sea urchin *P. lividus* are edible, and the rest of the body parts, usually discarded as waste, contain mutable connective tissue which can be exploited for the extraction of collagen (Barbaglio et al. 2015). The compass depressor ligaments and peristomial membranes in sea urchins are mutable collagenous structures. The compass depressor ligaments stabilize the position of Aristotle's lantern, which is the dental apparatus in sea urchins, and the peristomial membrane forms part of the body wall supporting the dental apparatus and connecting it to the hard calcified test. The compass depressor ligaments and peristomial membrane are rich in collagen fibrils (Barbaglio et al. 2015). The body walls of sea urchins and starfishes show highly packed collagen fibrils and conspicuous calcareous ossicles. On the contrary, sea cucumber body walls have loosely packed collagen fibrils, which are also homogenously distributed and have only small calcareous spicules. Because of this difference in the structural organization of collagen fibrils, stronger treatments such as disulfide bond disruption are required for the extraction of collagen from sea urchins and starfishes, whereas collagen from sea cucumbers can be extracted with mild non-denaturing methods (Ferrario et al. 2017).

The conventional methods of collagen extraction include the salting out, alkali-assisted, acid-assisted and enzyme-assisted extraction methods (Table 6.1) (Eser & Gozde 2021; Xu et al. 2021). Acid-assisted extraction is done using organic acids such as acetic acid, citric acid or lactic acid or using inorganic acids such as hydrochloric acid. Organic acids such as acetic acid and lactic acid

TABLE 6.1
Marine Sources, Extraction, Applications and Beneficial Properties of Collagen

Source Organism	Source Tissue	Extraction Condition	Applications and Beneficial Properties	Reference
Paracentrotus lividus	Peristomial membranes	Hypotonic buffer (10 mM Tris, 0.1% EDTA), decellularizing solution 10 mM Tris, 0.1% sodium dodecyl sulfate, phosphate-buffered saline, disaggregating solution (0.5 M NaCl, 0.1 M Tris-HCl pH 8.0, 0.1 M β-mercaptoethanol, 0.05 M EDTA-Na), dialysis against 0.5 M EDTA-Na solution (pH 8.0) and against dH$_2$O	Collagen membrane with higher stiffness and tensile strength than commercial collagen membrane	(Ferrario et al. 2017)
Echinaster sepositus	Aboral arm walls	Hypotonic buffer (10 mM Tris, 0.1% EDTA), decellularizing solution (10 mM Tris, 0.1% sodium dodecyl sulphate), phosphate-buffered saline, 1 mM citric acid (pH 3–4), disaggregating solution (0.5 M NaCl, 0.1 M Tris-HCl pH 8.0, 0.1 M β-mercaptoethanol, 0.05 M EDTA-Na), dialysis against 0.5 M EDTA-Na solution (pH 8.0) and against dH$_2$O	Collagen membrane with higher stiffness and tensile strength than commercial collagen membrane	(Ferrario et al. 2017)
Holothuria tubulosa	Whole body walls	Phosphate-buffered saline and gentamicin (40 mg/mL) for 5 days in stirring condition, filtration to obtain collagen suspension	Collagen membrane with higher stiffness and tensile strength than commercial collagen membrane	(Ferrario et al. 2017)
Paracentrotus lividus	Compass depressor ligament and Peristomial membrane	Disaggregating solution of 0.5 M NaCl, 0.05 M EDTA-Na, 0.1 MTris–HCl buffer (pH 8.0) and 0.2 M β-mercaptoethanol (1:20 wet w/v), centrifugation and re-suspension in 0.5 M EDTA-Na (pH 8.0) and distilled water	Mechanically adaptable biomaterial	(Barbaglio et al. 2015)
Tilapia	Skin	0.1 M NaOH (1–2 days), soaking in 0.5–1 M acetic acid (4–8 h), 0.1–0.5% pepsin (24–48 h), 0.4 M ammonium sulfate precipitation, dissolution in 0.5–1 M acetic acid, dialysis, and lyophilization	Collagen sponge and nanofibers for wound healing and skin regeneration	(Zhou et al. 2016)

(Continued)

TABLE 6.1 (CONTINUED)

Source Organism	Source Tissue	Extraction Condition	Applications and Beneficial Properties	Reference
Mixture of different species of sharks, mullet, guitarfish, weakfish, snapper, ray, squid, sea bass and pompano dolphin fish	Skins, heads, and skeletons	0.5 M acetic acid (24 h, 1:10 w/v, 25°C), papain (10 mg/kg, 1:2 w/v, 1 h, 24°C, pH 8–8.2), bleaching with NaClO (0.5%, 1:3 w/v, 30 min), ultra-filtration (30 kDa cellulose membrane), dissolution in HCl 0.5 M (4% solution), dialysis and freeze drying	Collagen hydrolysate with antioxidant and anti-microbial action, high foaming capacity, foaming stability and emulsion stability index	(Zamorano-Apodaca et al. 2020)
Marine Eel	Skin	0.5 M acetic acid (4°C, 42 h), precipitation in 0.9 M NaCl, suspension in 0.5 M acetic acid containing 1% pepsin (4°C, 24 h), salting out using NaCl, dissolution in 0.5 M acetic acid, dialysis against 0.1 M acetic acid and later with water at 4°C, freeze-drying and lyophilization of collagen	Alginate/collagen hydrogel scaffolds using extrusion-based 3D printing technology	(Govindharaj et al. 2019)

yield higher content of collagen, whereas citric acid and hydrochloric acid yield smaller amounts of collagen due to protein denaturation (Skierka & Sadowska 2007). Salting out using sodium chloride is another efficient method for the extraction of collagen.

Several biotechnological approaches have been developed for the extraction of collagen and other valuable biomolecules from seafood processing byproducts. Biotechnological approaches include mainly fermentative extraction and enzymatic extraction. In fermentative extraction, microorganisms grown on seafood byproducts produce acids or enzymes which then act simultaneously on the extracellular matrix of the seafood tissues, resulting in solubilization and detachment of biomolecules (Bruno et al. 2019). Enzymatic extraction of collagen is usually done using proteolytic enzymes such as pepsin, pancreatin and trypsin, which are of plant, animal or microbial origin (Eser & Gozde 2021). The acid dissolution of collagen can largely maintain the triple helix structure of collagen, whereas enzymatic extraction can reduce the antigenicity of collagen by removing the N-terminal and C-terminal antigenic regions of collagen peptides (Zhou et al. 2016). The final yield of collagen depends on the marine species used as source and the age of animal tissue used for extraction. It is also influenced by several other parameters such as the initial fresh or frozen condition of the tissue, pH and ionic strength of the extraction solvent, extraction temperature, sample-volume ratio, types of acids and enzyme used and time of exposure (Salvatore et al. 2020).

Innovative technologies are emerging to improve the yield and properties of the final products. In comparison to the conventional techniques, improved physical assisted extraction methods have the advantage of retaining higher molecular weight with peptide spectrum similar to that extracted with acid-assisted methods (Xu et al. 2021). Physical assisted extraction techniques such as supercritical fluid extraction and deep eutectic solvent extraction are innovative green technologies with simple and sustainable operations and propitious yield of collagen on a large scale (Liu et al. 2022). Supercritical fluid extraction is an advanced separation technique which makes use of fluids with an elevated critical point of temperature. Supercritical fluids enhance the extraction of proteins by penetrating deeper and faster into solid matrices (Abhari & Khaneghah 2020). Water acidified with carbon dioxide is most commonly used for the extraction of collagen from marine sponges (Silva et al. 2016) since it prevents protein denaturation by providing a non-oxidizing atmosphere (Abhari & Khaneghah 2020). The extraction efficiency of supercritical fluids can be further enhanced by incorporating an organic solvent such as methanol or ethanol. The extraction of collagen from the marine demosponge *Chondrosia reniformis* using water acidified with carbon dioxide yielded a product of high purity (Silva et al. 2016). The biocompatibility of collagen extracted using water acidified with carbon dioxide in cytotoxic evaluation suggests the high industrial potential of this technique for the extraction of bioactive compounds for biomedical applications. Deep eutectic solvents are advantageous in that they are non-flammable, highly thermostable, less volatile and recyclable and have high solubilization potential for the extraction of compounds. Among the different deep eutectic solvents, choline chloride-oxalic acid is highly efficient for the extraction of collagen peptides of high purity from the skin tissues of fishes (Bai et al. 2017).

6.3 PROPERTIES OF MARINE COLLAGEN

6.3.1 Amino Acid Composition

The primary beneficial property that distinguishes marine collagen as an excellent biomaterial for biomedical applications is its remarkable biocompatibility, which is attributed to its amino acid composition that is highly similar to mammalian collagen (Salvatore et al. 2020). The collagen gene sequences are highly conserved and similar across species. This remarkable property of marine collagen gives it high biocompatibility, low immunogenicity and low cytotoxicity that is highly favorable for cell adhesion and proliferation and survival of cells in tissue engineering and wound healing applications (Choi et al. 2015; Liu et al. 2022).

Marine collagen, especially fish collagen, contains repetitive sequences of glycine and hydroxyproline (Figure 6.1). Alanine, proline, hydroxyproline and glutamic acid are also high in marine collagen. Glycine contributes to the formation of the α-helix and stability of the triple helix configuration of collagen. Marine collagen is also notable for its low levels of cysteine, tyrosine, histidine and hydroxylysine. Histidine is the precursor to histamines associated with allergic responses, and the low content of histidine is considered an index for the low antigenicity of marine collagen. The high contents of proline and hydroxyproline also strengthen the triple helix and contribute to thermal stability. Cysteine forms disulfide bridges between α-helices and stabilizes the collagen triple helix (Salvatore et al. 2020).

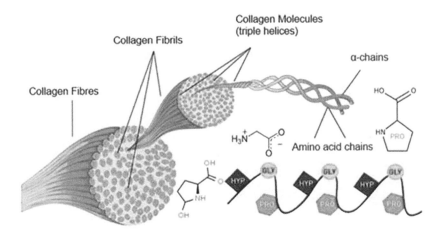

FIGURE 6.1 Triple helical structure of marine collagen protein (Rastogi et al. 2022) (https://creativecommons.org/licenses/by/4.0/).

6.3.2 MUTABLE NATURE

Collagen from echinoderms is proposed as useful for developing "smart dynamic biomaterials" for applications in tissue engineering and regenerative medicine because of their origin from a unique mutable connective tissue (Ferrario et al. 2017). The mutable property of echinoderm collagen is highly appreciable in the cosmetic industry for anti-ageing treatment since it facilitates the modification of mechanical properties. Moreover, in the pharmaceutical industry, drugs made using mutable collagen can alter the interaction of connective tissues with pathogens (Barbaglio et al. 2015).

6.3.3 TENSILE STRENGTH

The natural collagen fibrils obtained from echinoderms are also able to maintain their stable original structure when compared to mammalian collagen. This is advantageous since membranes and scaffolds prepared from the collagen of echinodermal origin can deliver excellent mechanical performance and can be applied for biomedical applications requiring highly resistant materials with three-dimensional fibril organization (Barbaglio et al. 2015). Echinoderm-derived collagen membranes offer the possibility of producing membranes which are much thinner but with a high tensile strength that can resist mechanical stress. This is beneficial in the manufacture of biomaterials since the higher the tensile strength, the better the ability to resist mechanical stress and rupture (Ferrario et al. 2017).

Echinoderm-derived collagen membranes from sea urchins, starfish and sea cucumbers show significantly higher stiffness and tensile strength in comparison to commercial bovine collagen membranes (Ferrario et al. 2017). Moreover, echinoderm-derived collagen fibrils have inherent decorations of glycosaminoglycan on their surface (Figure 6.2), which is highly significant in maintaining the integrity of fibrils. It is also pertinent in tissue engineering applications for the migration, adhesion, proliferation and differentiation of cells (Ferrario et al. 2017). This is advantageous over bovine collagen membranes, which require the addition of glycosaminoglycan for improved performance in tissue engineering applications.

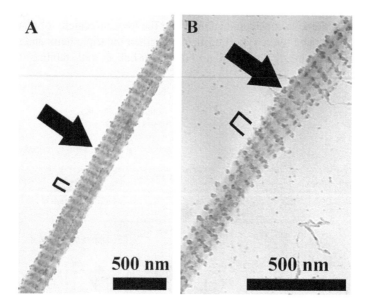

FIGURE 6.2 Transmission electron micrograph showing glycosaminoglycan (GAG) distribution (arrows) on echinoderm collagen fibril surface according to D-patterning (square brackets). a) Starfish-derived collagen fibril; b) sea cucumber-derived collagen fibril.
Source: (Reprinted from Ferrario et al. (2017) with permission from Elsevier).

6.3.4 POROSITY

Moreover, echinoderm-derived collagen membranes have superficial porosity that is much smaller than that human cells. This renders a more highly efficient barrier effect than commercial bovine-derived collagen membranes. This is a very important property required for guided tissue regeneration, wherein proper compartmentalization is needed between two different anatomical parts to prevent the mixture of adjacent regenerating tissues and to facilitate the post-surgical healing process (Ferrario et al. 2017). In addition, the porosity and three-dimensional structure of collagen membranes developed from echinoderm mutable connective tissues can be modified as required. Collagen scaffolds developed using marine collagen extracted from coelenterates and fishes also have high porosity with interconnected pore structures (Choi et al. 2015).

6.4 BIOLOGICAL ACTION AND HEALTH BENEFITS OF MARINE COLLAGEN

The low-molecular weight peptides obtained by the hydrolysis of fish collagen demonstrate higher biological actions than native high-molecular weight collagen because of the increased accessibility of amino acids. The biological actions displayed by the peptides of fish collagen hydrolysate include antioxidant, antimicrobial, angiotensin-conversion, anti-hyperglycemic, anti-Alzheimer's and neuroprotective action (Zamorano-Apodaca et al. 2020).

6.4.1 ANTIOXIDANT ACTION

The low-molecular peptides in fish collagen hydrolysate and squid collagen hydrolysate exhibit antioxidant activity against 1,1-diphenyl-2-picrylhydrazyl radical (DPPH) and hydroxyl radical (OH). The amino acids capable of donating electron/hydrogen such as glycine, glutamic acid, alanine and aspartic acid are better exposed in low-molecular weight peptides facilitating better interaction with the lipids and free radical neutralizing them into stable products. The antioxidant property of marine collagen is attributed to the high content of hydrophobic amino acids with the ability to chelate pro-oxidative transition metals by donating electrons/hydrogen (Zamorano-Apodaca et al. 2020). The antioxidant peptides from marine collagen hydrolysate can be useful in nutraceutical and pharmaceutical products.

6.4.2 ANTIBACTERIAL ACTION

Marine collagen contains rich amounts of hydrophobic amino acids such as alanine, leucine, isoleucine, valine, proline, phenylalanine, methionine, serine, tyrosine and glycine and positively charged amino acids such as lysine, arginine and histidine. The positively charged amino acids of marine collagen hydrolysate can interact with the negatively charged groups on the bacterial cell wall polysaccharides and cell membrane phospholipids, thus penetrating the cell membranes, ultimately leading to cell lysis (Zamorano-Apodaca et al. 2020). The antibacterial action of cationic collagen-derived peptides can be modulated by modifying their net charge and amount of hydrophobic amino acids in the peptide sequence. The methionine present in the collagen peptide sequence can inhibit the replication of bacterial DNA. The molecular size and the amino acid composition of collagen-derived peptides are decisive in their antimicrobial action. The smaller the molecular size, the higher the antimicrobial action of collagen (Zamorano-Apodaca et al. 2020). Hence, by modulating the degree of hydrolysis of collagen, it is possible to generate different collagen-derived peptides with a wide spectrum of antimicrobial action.

6.4.3 ANTI-OBESOGENIC AND ANTI-DIABETIC ACTION

Marine collagen from fish tissue hydrolysate is functional as an anti-diabetic peptide since they have the potential to increase the cellular uptake of glucose by increasing the sensitivity of target cells to the hormone insulin (Abhari & Khaneghah 2020). The anti-obesogenic effect of marine collagen is exerted through the regulation of adipogenic differentiation and lipid metabolism (Lee et al. 2017). Collagen peptides can inhibit the differentiation of new adipocytes from mesenchymal stem cells by regulating the expression of adipogenic genes and transcription factors (Tometsuka et al. 2017). Marine collagen peptides upregulate the biosynthesis of unsaturated fatty acids and cellular metabolism of fatty acids and also modulate the insulin sensitivity of cells (Zhu et al. 2017). Administration of collagen also decreases the serum levels of cholesterol, triglycerides and low-density lipoproteins, at the same time increasing the levels of high-density of lipoproteins. Marine collagen peptides exert a dose-dependent hypoglycemic effect in reducing the serum level of glucose (Salvatore et al. 2020).

6.4.4 Food Applications

Marine collagen extracted from the skin, bones, scales and fins of fishes is used as food emulsion to modify the gelation and maintain the texture of foods. Collagen fibers have high tensile strength and have application in the packaging of foods. Collagen from fish wastes is also used for the manufacture of chewable tablets with good texture (Nawaz et al. 2020). Collagen can be transformed into gelatin through thermal and chemical treatments by breaking the hydrogen bonds that stabilize the triple helix. Gelatin derived from collagen can prevent oxidation of lipids, preserve the taste of foods and increase their shelf life. Hence gelatin is used as a consistence enhancer and food stabilizer. Collagen peptides can improve health and increase longevity because of their ability to increase bone formation and bone mineral density and alleviate osteoarthritis. Hence hydrolyzed collagen is used to enrich functional drinks. There is some evidence that the administration of collagen peptides can improve bone mineral density in post-menopausal women (Salvatore et al. 2020).

6.4.5 Cosmetic Applications

Marine collagen can enhance the hydration and elasticity of the skin, minimize the formation of wrinkles and repair collagen and elastin fibers in the skin that are damaged on exposure to light. The natural moisturizing, softening and glowing properties also render marine collagen a suitable active ingredient in anti-ageing formulations for topical application (Salvatore et al. 2020). In comparison to high molecular weight native collagen, low molecular weight collagen-derived peptides give better results in cosmetic formulations because of their better solubility and water binding properties and ease of penetration into the skin (Alves et al. 2017). Marine collagen-derived peptides in cosmetic formulations can increase the action of free-radical scavenging endogenous antioxidant enzymes, thereby reducing the oxidative stress-induced skin ageing process.

6.5 BIOMEDICAL APPLICATIONS OF MARINE COLLAGEN

Marine collagen finds excellent applications as a biomaterial in tissue engineering and regenerative medicine owing to its biocompatibility. The biomimetic nature of collagen facilitates the proliferation and differentiation of cells in scaffolds for tissue engineering.

6.5.1 Scaffolds for Tissue Engineering and Regenerative Medicine

The scaffold ideal for tissue engineering is one that can mimic the native extracellular matrix and thus facilitate cell adhesion, proliferation, differentiation and migration. Scaffolds made out of electrospun marine collagen nanofibers are porous, resemble the native extracellular matrix and are hence suitable for 3D cell culture and tissue engineering applications. Fish collagen scaffolds can promote cell adhesion by promoting the expression of vinculin and E-cadherin (Choi et al. 2015). Vinculin is a membrane cytoskeletal protein in focal adhesion plaques, and E-cadherin is a calcium-dependent cell–cell adhesion molecule. Hence fish collagen scaffolds aid the adhesion of epithelial cells to scaffolds by providing chemically and biologically favorable cues for their growth and proliferation. Furthermore, fish collagen scaffolds also promote the phosphorylation and activation of focal adhesion kinase, which is a protein tyrosine kinase. The activation of focal adhesion kinase can initiate a cascade of reactions via numerous signal transduction pathways that eventually set in motion the increased proliferation and survival of epithelial cells in the scaffold (Choi et al. 2015). Fish collagen scaffolds also promote the enhanced expression of IL-7, a thymopoietic protein, and cytokeratin-8, a cytoskeletal protein, which are key indicators of the physiological function of cortical reticular epithelial cells. All these factors contribute to better performance of fish collagen as an excellent biomaterial scaffold for tissue engineering in comparison to scaffolds made out of synthetic materials.

Collagen from echinoderms is useful for developing collagen-barrier membranes for guided tissue regeneration.

6.5.2 Wound Healing

Wound healing is a complex physiological process involving inflammatory responses, formation and differentiation of new epithelial cells, deposition of collagen and vascularization of newly formed skin tissues. Wound dressings that induce tissue regeneration can accelerate the process of wound healing (Zhou et al. 2016). Keratinocytes, one of the chief cells involved in wound healing, undergo proliferation and differentiation to form epithelial cells. Collagen fibers extracted from fish skin tissues can act as biomimetic materials that mimic the structure and function of the native extracellular matrix and facilitate the adhesion and proliferation of keratinocytes. They can also induce keratinocytes to differentiate into epidermal tissues by the upregulation of related genes such as genes for involucrin, filaggrin and transglutaminase1 (TGase1) (Choi et al. 2015; Zhou et al. 2016). TGase1 is the chief enzyme involved in the synthesis of the cornified layer of the skin, whereas involucrin and filaggrin are expressed in the suprabasal epidermal layer, and these are required for the formation of an integrated epidermis (Zhou et al. 2016).

Alleviating the inflammatory response is very pertinent in wound healing. Marine collagen exerts anti-inflammatory action by inhibiting the expression of proinflammatory cytokines such as tumor necrosis factor-α (TNF-α), interleukins and inducible nitric oxide synthase (Salvatore et al. 2020).

6.6 SUMMARY AND CONCLUSIONS

Marine collagen obtained from marine animal species belonging to different groups such as poriferans, coelenterates, molluscs, echinoderms and fishes is acclaimed as an excellent and sustainable alternative to bovine and porcine collagen for a wide range of applications in the food, nutraceutical and pharmaceutical sectors as well as in tissue engineering and regenerative medicine. Collagens have been extracted and characterized from a wide range of marine organisms. Collagen from echinoderms is unique in terms of its mechanical adaptability and mutable nature because of its origin from mutable connective tissue. Though the beneficial properties of marine collagen position it as a sustainable alternative to mammalian collagen, scientific research is needed to validate its immunogenicity, biocompatibility and biodegradability for application in human subjects. Intensive research is needed to develop collagen derived from echinoderms as a smart biomaterial for tissue engineering and regenerative medicine. Not only has marine collagen opened up the opportunity for the valorization of marine wastes, but it also offers chances of resolving the environmental threats posed by the "dangerous" species of invertebrates attacking the coral reefs (e.g. coral feeder starfish *Acanthaster planci*). The increased market demand for marine collagen together with concerns about declining wild populations of marine species has raised the need for aquaculture practices as an effective alternative to capture fisheries. The increasing market demand for fish and marine invertebrate collagen and their derived peptides suggests a progressive trend in its application in the development of medical devices for tissue engineering and regenerative medicine.

REFERENCES

Abhari, K., Khaneghah, A.M., 2020. Alternative extraction techniques to obtain, isolate and purify proteins and bioactive from aquaculture and by-products, in: Advances in Food and Nutrition Research. Elsevier Inc., pp. 35–52. https://doi.org/10.1016/bs.afnr.2019.12.004

Alves, A.L., Marques, A.L.P., Martins, E., Silva, T.H., Reis, R.L., 2017. Cosmetic potential of marine fish skin collagen. Cosmetics 4 (39), 1–16. https://doi.org/10.3390/cosmetics4040039

Bai, C., Wei, Q., Ren, X., 2017. Selective extraction of collagen peptides with high purity from cod skins by deep eutectic solvents. ACS Sustain. Chem. Eng. 5, 7220–7227. https://doi.org/https://doi.org/10.1021/acssuschemeng.7b01439

Barbaglio, A., Tricarico, S., Ribeiro, A.R., Di Benedetto, C., Barbato, M., Dessì, D., Fugnanesi, V., Magni, S., Mosca, F., Sugni, M., Bonasoro, F., Barbosa, M.A., Wilkie, I.C., Candia Carnevali, M.D., 2015. Ultrastructural and biochemical characterization of mechanically adaptable collagenous structures in the edible sea urchin *Paracentrotus lividus*. Zoology 118, 147–160. https://doi.org/10.1016/j.zool.2014.10.003

Bruno, S.F., Ekorong, F.J.A.A., Karkal, S.S., Cathrine, M.S.B., Kudre, T.G., 2019. Green and innovative techniques for recovery of valuable compounds from seafood by-products and discards: A review. Trends Food Sci. Technol. 85, 10–22. https://doi.org/10.1016/j.tifs.2018.12.004

Choi, D.J., Choi, S.M., Kang, H.Y., Min, H.-J., Lee, R., Ikram, M., Subhan, F., Jin, S.W., Jeong, Y.H., Kwak, J.-Y., Yoon, S., 2015. Bioactive fish collagen/polycaprolactone composite nanofibrous scaffolds fabricated by electrospinning for 3D cell culture. J. Biotechnol. 205, 47–58. https://doi.org/10.1016/j.jbiotec.2015.01.017

Dario Fassini, Iain C. Wilkie, Marina Pozzolini, Cinzia Ferrario, Michela Sugni, Miguel S. Rocha, Marco Giovine, Francesco Bonasoro, Tiago H. Silva, and Rui L. Reis, 2021. Biomacromolecules 2021 22 (5), 1815–1834. https://doi.org/https://doi.org/10.1021/acs.biomac.1c00013

Eser, B.E., Gozde, K.I., 2021. Marine collagen. Stud. Nat. Prod. Chem. 71, 121–139. https://doi.org/10.1016/B978-0-323-91095-8.00002-7

Ferrario, C., Leggio, L., Leone, R., Di Benedetto, C., Guidetti, L., Coccè, V., Ascagni, M., Bonasoro, F., La Porta, C.A.M., Candia Carnevali, M.D., Sugni, M., 2017. Marine-derived collagen biomaterials from echinoderm connective tissues. Mar. Environ. Res. 128, 46–57. https://doi.org/10.1016/j.marenvres.2016.03.007

Govindharaj, M., Roopavath, U.K., Rath, S.N., 2019. Valorization of discarded marine eel fish skin for collagen extraction as a 3D printable blue biomaterial for tissue engineering. J. Clean. Prod. 230, 412–419. https://doi.org/10.1016/j.jclepro.2019.05.082

Lee, E.J., Hur, J., Ham, S.A., Jo, Y., Lee, S.Y., Choi, M.-J., Seo, H.G., 2017. Fish collagen peptide inhibits the adipogenic differentiation of preadipocytes and ameliorates obesity in high fat diet-fed mice. Int. J. Biol. Macromol. 104, Part, 281–286. https://doi.org/https://doi.org/10.1016/j.ijbiomac.2017.05.151

Liu, S., Lau, C.S., Liang, K., Wen, F., Teoh, S.H., 2022. Marine collagen scaffolds in tissue engineering. Curr. Opin. Biotechnol. 74, 92–103. https://doi.org/10.1016/j.copbio.2021.10.011

Nawaz, A., Li, E., Irshad, S., Xiong, Z., Xiong, H., 2020. Valorization of fisheries by-products: Challenges and technical concerns to food industry. Trends Food Sci. Technol. 99, 34–43. https://doi.org/10.1016/j.tifs.2020.02.022

Rastogi, K., Vashishtha, R., Shaloo, Dan, S., 2022. Scientific advances and pharmacological applications of marine derived-collagen and chitosan. Biointerface Res. Appl. Chem. 12, 3540–3558. https://doi.org/10.33263/BRIAC123.35403558

Salvatore, L., Gallo, N., Natali, M.L., Campa, L., Lunetti, P., Madaghiele, M., Blasi, F.S., Corallo, A., Capobianco, L., Sannino, A., 2020. Marine collagen and its derivatives: Versatile and sustainable bioresources for healthcare. Mater. Sci. Eng. C 113, 110963. https://doi.org/10.1016/j.msec.2020.110963

Silva, J., Barros, A., Aroso, I., Fassini, D., Silva, T., Reis, R., Duarte, A., 2016. Extraction of collagen/gelatin from the marine demosponge *Chondrosia reniformis* (Nardo, 1847) using water acidified with carbon dioxide—process optimization. Ind. Eng. Chem. Res. 55, 6922–6930. https://doi.org/https://doi.org/10.1021/acs.iecr.6b00523

Skierka, E., Sadowska, M., 2007. The influence of different acids and pepsin on the extractability of collagen from the skin of Baltic cod (*Gadus morhua*). Food Chem. 105, 1302–1306. https://doi.org/https://doi.org/10.1016/j.foodchem.2007.04.030

Steinke, J.W., Platts-Mills, T.A.E., Commins, S.P., 2015. The alpha-gal story: Lessons learned from connecting the dots. J. Allergy Clin. Immunol. 135, 589–596. https://doi.org/10.1016/J.JACI.2014.12.1947

Tometsuka, C., Koyama, Y., Ishijima, T., Toyoda, T., Teranishi, M., Takehana, K., Abe, K., Nakai, Y., 2017. Collagen peptide ingestion alters lipid metabolism-related gene expressionand the unfolded protein response in mouse liver. Br. J. Nutr. 117, 1–11. https://doi.org/10.1017/S0007114516004384

Xu, N., Peng, X.-L., Li, H.-R., Liu, J.-X., Cheng, J.-S.-Y., Qi, X.-Y., Ye, S.-J., Gong, H.-L., Zhao, X.-H., Yu, J., Xu, G., Wei, D.-X., 2021. Marine-derived collagen as biomaterials for human health. Front. Nutr. 8, 702108. https://doi.org/https://doi.org/10.3389/fnut.2021.702108

Yamada, S., Yamamoto, K., Ikeda, T., Yanagiguchi, K., Hayashi, Y., 2014. Potency of fish collagen as a scaffold for regenerative medicine. Biomed Res. Int. 2014, 302932. https://doi.org/10.1155/2014/302932

Yamamoto, K., Igawa, K., Sugimoto, K., Yoshizawa, Y., Yanagiguchi, K., Ikeda, T., Yamada, S., Hayashi, Y., 2014. Biological safety of fish (tilapia) collagen. Biomed Res. Int. 2014, 630757. https://doi.org/10.1155/2014/630757

Zamorano-Apodaca, J.C., García-Sifuentes, C.O., Carvajal-Millán, E., Vallejo-Galland, B., Scheuren-Acevedo, S.M., Lugo-Sánchez, M.E., 2020. Biological and functional properties of peptide fractions obtained from collagen hydrolysate derived from mixed by-products of different fish species. Food Chem. 331, 127350. https://doi.org/10.1016/j.foodchem.2020.127350

Zhou, T., Wang, N., Xue, Y., Ding, T., Liu, X., Mo, X., 2016. Electrospun tilapia collagen nanofibers accelerating wound healing via inducing keratinocytes proliferation and differentiation. Colloids Surfaces B Biointerfaces 143, 415–422. https://doi.org/10.1016/j.colsurfb.2016.03.052

Zhu, C.F., Zhang, W., Mu, B., Zhang, F., Lai, N., Zhou, J.X., Xu, A.M., Liu, J.G., Li, Y., 2017. Effects of marine collagen peptides on glucose metabolism and insulin resistance in type 2 diabetic rats. J. Food Sci. Technol. 54, 2260–2269. https://doi.org/https://doi.org/10.1007/s13197-017-2663-z

7 Application of Functional Aquafeed in Sustainable Aquaculture

Yu-Hung Lin, Winton Cheng and Hsin-Wei Kuo

CONTENTS

7.1 Introduction 89
7.2 Functional Feed Additives in Plant-Based Aquafeeds 90
    7.2.1 Taurine 90
    7.2.2 Cholesterol 91
    7.2.3 Nucleotides 92
    7.2.4 Organic Acid 92
    7.2.5 Selenium 92
7.3 Antibiotic Free or Reduced Aquafeed 93
    7.3.1 Essential Nutrients 93
        7.3.1.1 Amino Acids 93
        7.3.1.2 Lipids and Fatty Acids 93
        7.3.1.3 Vitamins and Minerals 93
    7.3.2 Other Immunostimulants 95
        7.3.2.1 Seaweed Polysaccharides 96
        7.3.2.2 Water Hyacinth 96
        7.3.2.3 Banana Peel 96
        7.3.2.4 Cacao Pod Husks 96
        7.3.2.5 Lemon Peel 97
7.4 Conclusions and Perspectives 98
References 98

7.1 INTRODUCTION

The world population, currently 7.9 billion, is projected to exceed 9 billion by 2050 (UN Division of Population, https://population.un.org/). A growing world population leads to an increasing demand for food. Additionally, the global climate is changing, making land the focus of intensified competition as resources for a variety of uses, in turn decreasing agricultural and livestock production. Marine resources are important to fill the gap in food supply. Capture and aquaculture fisheries can satisfy the growing demands of global communities for more fish products in human diets. Statistical data from the Food Agriculture Organization (FAO) show that since the 1990s, the greatest growth in production has been from aquaculture, while the production of capture fisheries has been relatively stable (FAO, www.fao.org/). A continuation of sustainable aquaculture development will be needed to maintain this upward trend and meet future demand.

    The success of the aquaculture industry depends on high-quality aquafeed. Quite different from land animals, the protein requirements for aquatic animals are usually high, and the dietary protein is mainly dependent on marine-derived protein, such as fish meal (NRC, 2011). Since the 1980s, several feedstuffs from plant, animal by-products, or single-cell proteins have been well established to reduce the use of fish meal. Non-marine protein sources, especially soybean meal, have become important feedstuffs due to the shortage of production and the high cost of fish meal around the world. Our previous studies indicated that several negative impacts, such as suppression of immunity, nutrient digestibility, and metabolism, as well as enteritis and fatty liver, were observed in fish and shrimp fed plant protein–based diets (Lin & Cheng, 2017; Lin & Mui, 2017; Lin & Lu, 2020; Wu et al., 2020; Yong et al., 2020). These could lead to poor feeding conversion and health status and high water pollution. Specific feed formulations or functional feed additives have been expected to improve these negative effects on fish and shrimp.

Furthermore, another important issue in sustainable aquaculture is the use of antibiotics. Aquaculture farmers are interested in developing cost-effective preventive measures that can limit the spread of or reduce diseases. This has often led farmers to use antibiotics excessively as a prophylactic in an attempt to mitigate these challenges. Supplementing antibiotics in feed is one of the most effective means of providing medication to fish to treat bacterial infections, especially in large culture systems. Controlling the amount of antibiotics fed to cultured organisms is critical. Proper feeding rates and withdrawal times must be followed to reduce antibiotic deposition in fish tissues or release into rearing water that may be discharged into the aquatic environment. Dietary immunostimulants are a useful strategy for disease control in the modern aquaculture industry. Immunostimulants, such as bacterial derivatives, polysaccharides, animal and plant extracts, synthetic chemicals, hormones, cytokines, and nutritional factors have been well documented in fish and shrimp nutrition. Of these, nutritional factors and plant extracts are commonly acceptable in aquafeed. Therefore, this chapter reviews the nutritional strategy in plant-based aquafeed, as well as strategies in health control for fish and shrimp.

7.2 FUNCTIONAL FEED ADDITIVES IN PLANT-BASED AQUAFEEDS

As mentioned previously, in aquafeed, less expensive and more sustainable protein sources substituted for fish meal are required. Since the 1980s, many studies have successfully identified many kinds of alternative protein sources to replace the use of fish meal (NRC, 2011). For example, (1) plant protein sources, such as soybean meal, canola meal, sunflower meal, cottonseed meal, peanut cake, palm kernel meal, and lupin meal; (2) animal protein sources, such as meat and bone meal, poultry meal, blood meal, feather meal, shrimp head meal, and crab meal; and (3) other protein sources, such as single-cell protein from microalgae or bacteria and insect meal (NRC, 2011). Among these feedstuffs, soybean protein is the most widely used feed ingredient. Previous studies found that giant grouper (*Epinephelus lanceolatus*) fed a plant protein-based diet showed poor growth, low nutrient digestibility (Lin & Cheng, 2017; Lin & Lu, 2020), high oxidative stress (Lin & Cheng, 2017; Wu et al., 2020; Yong et al., 2020), enteritis (Lin & Cheng, 2017), high visceral fat, and fatty liver (Yong et al., 2020). In a study of white shrimp (*Litopenaeus vannamei*), ingestion of a diet containing a large amount of plant protein caused low immune responses and resistance to disease, as well as morphological damage to the hepatopancreas (Lin & Mui, 2017; Lin & Chen, 2022). These negative impacts are suggested to be due to taurine, cholesterol, or nucleotide deficiency and the presence of antinutritional factors. Furthermore, some short-chain fatty acids (organic acids) were also reported to repair intestinal damage or improve nutrient digestibility.

7.2.1 TAURINE

Taurine is a neutral β-amino acid analog derived from the metabolism of sulfur-containing amino acids. In mammals or fish, taurine has already been known to play an important physiological role, including membrane stabilization, antioxidation, detoxification, osmoregulation, and conjugation with bile acids (NRC, 2011). In the past, taurine has been considered a nonessential nutrient because taurine content is high in marine ingredients and has been shown to be synthesized *de novo* through the transulfuration pathway of methionine in fish (reviewed by Salze & Davis, 2015). However, a typical taurine deficiency, green liver syndrome, has been demonstrated in marine fish fed a diet rich in plant feedstuffs (Salze & Davis, 2015). The phenomenon is possibly due to impaired liver bile pigment excretion and hemolytic biliverdin overproduction linked to dietary taurine deficiency. Our recent study indicated that taurine supplementation in a plant protein-based diet exhibited improvement in growth performance and physiological responses for giant grouper (*Epinephelus lanceolatus*) (Lin & Lu, 2020). It suggests that taurine may be a semi- (or conditional) essential nutrient for marine fish under special situations like plant protein-based feed.

Normally, taurine essentiality is largely associated with the activity of key enzymes responsible for sulfur amino acid metabolism (Salze & Davis, 2015). The cysteinesulfinate-dependent pathway is considered the main pathway for taurine biosynthesis in mammals and fish. Sulfur-containing amino acids, such as methionine or cysteine, are bio-converted to taurine, which is highly regulated by cysteine dioxygenase (CDO) and cysteinesulfinate decarboxylase (CSD), respectively. Giant grouper fed a high soybean meal diet (33.3%) showed downregulation of taurine-synthesizing enzymes, CDO and CSD, and gene expression in the liver compared to giant grouper fed an entirely fish meal diet (Lin, 2019). This suggests that plant feedstuff included in diets may adversely affect biosynthesis for the species. Our study found liver CSD gene expression was induced by dietary 0.5% methionine inclusion, while tissue taurine concentration was similar among all groups (Lin, 2019). The results indicate the possibility that giant grouper has taurine biosynthesis capability, but it may not be sufficient for its endogenous requirement.

The apparent lipid digestibility was negatively affected by replacing dietary fish meal with soybean meal giant grouper (Lin & Cheng, 2017). The bile juice involved with taurine may be the reason for the influence of lipid digestibility for fish. Our recent study indicated that giant grouper fed a soybean meal-based diet depressed the apparent lipid digestibility, and 0.1% taurine supplementation in the diet can significantly enhance the lipid digestibility of the fish (Lin & Lu, 2020).

7.2.2 Cholesterol

Cholesterol is a key sterol that serves as a precursor to physiologically active compounds, including sex hormones, molting hormone, adrenal corticoids, bile acids, and vitamin D in aquatic animals (NRC, 2011). This nutrient is considered a non-essential nutrient for fish but an essential nutrient for crustaceans relative to the ability to *de novo* synthesize cholesterol. However, hypocholesterolemic effects have been reported in fish and shrimp (Wu et al., 2020; Lin et al., 2017a) fed a soybean meal-based diet. This phenomenon may be associated with deficiency of cholesterol or the presence of antinutritional factors such as saponin, phytosterol, or non-starch polysaccharides in sources of soy protein.

Interestingly, giant grouper fed a diet with high levels of soybean meal showed low plasma cholesterol concentrations but high hepatic cholesterol concentration (Wu et al., 2020). The study also found the up-regulation of the hepatic 3-hydroxy-3-methylglutaryl coenzyme A (HMG-CoA) reductase, a rate-limiting enzyme for cholesterol synthesis, gene expression in fish fed the soybean meal-based diet compared with the fish fed diet with fish meal. In addition, primary bile acids are biosynthesized by cholesterol in the liver, and the pathway is mainly regulated by the key enzyme cholesterol-7α hydroxylase (CYP7A1). In this grouper study, the gene expression of hepatic CYP7A1 was not significantly upregulated when fish were fed a soybean meal-based diet (Wu et al., 2020). Normally, excess cholesterol could initiate the expression of the liver CYP7A1 gene to convert cholesterol to bile salts and excrete it by the fish. However, grouper fed plant protein-based diets didn't increase the conversion of cholesterol to bile acids. This was possibly because high plant protein ingestion injured the normal liver function for the fish. Deteriorated histological changes and high oxidative stress in the liver have been found in another study of grouper (Yong et al., 2020) fed a high soybean meal–based diet, possibly supporting this suggestion.

The cholesterol concentrations in the hepatopancreas and hemolymph of Pacific white shrimp also decreased linearly with increasing levels of soybean meal inclusion (Lin et al., 2017a). This study found that the expressions of the relative molting gene, the ecdysteroid receptor (EcR) and the retinoic-X receptor (RXR), were also negatively regulated by shrimp fed a diet with high levels of soybean meal. Unlike in fish, cholesterol is necessary for shrimp to maintain normal growth and molting signal gene expressions when shrimp are fed a plant protein–based diet.

7.2.3 Nucleotides

Nucleotides have essential physiological and biochemical functions, including encoding and deciphering genetic information; mediating energy metabolism and cell signaling; and serving as components of coenzymes, allosteric effectors, and cellular agonists in terrestrial animals. Because of the active *de novo* synthesis of nucleotides, mainly in the liver, most animals appear to be almost independent of exogenous nucleotides.

It should be noted that the nucleotide concentration in soybean meal is approximately only half that in fish meal. A previous study indicated that hybrid tilapia (*Oreochromis niloticus* × *O. aureus*) fed a diet with 120–240 mg/kg nucleotides in a high–soybean meal diet showed higher nonspecific immune responses and resistance to the *Streptococcus iniae* pathogen (Shiau et al., 2015). A similar finding was reported in a white shrimp study (Shiau & Lin, 2015): immune responses and resistance to *Vibrio alginolyticus* pathogen were improved in shrimp fed diets with 60–120 mg/kg of nucleotides with feed based on plant protein. These two studies clearly suggest that the essentiality of nucleotides should be carefully considered in diets containing high levels of plant protein.

7.2.4 Organic Acid

Soybean meal included at high levels has been reported to induce some damage to the intestinal microvilli in giant grouper (Lin & Cheng, 2017; Yong et al., 2020). This phenomenon is well documented in many species and is defined as "soybean meal-induced enteritis."

Organic acids are organic carboxylic compounds that have the general structural formula R–COOH, and their acidity is associated with the carboxyl group (–COOH) (Lim et al., 2015). Feed supplementation with organic acids has been shown to lead to a lower duodenal pH, improved nitrogen retention, and increased nutrient digestibility. Supplementation of butyric acid and lactic acid in the diet shows several benefits for giant grouper (Lin & Cheng, 2017; Yong et al., 2020). Butyrate and lactic acid supplemented in the diet at 1% can improve nutrient digestibility and intestinal morphology in fish, excluding growth performance. In two studies, liver oxidative stress seemed to be induced by the inclusion of soybean meal in the grouper diet. This oxidative stress was also recovered by organic acid dietary supplementation.

7.2.5 Selenium

Selenium (Se) derived from a fish dietary source is considered to be more readily utilized than that derived from a plant source because Se concentration in plants is rather low and is largely dependent on soil Se concentration. Plants are capable of converting the Se ion to selenoamino acids, such as Se-methyl-selenocysteine and γ-glutamyl-Se-methyl-selenocysteine, and are generally presented as Se-methyl-selenocysteine, which represented about 66% of selenoamino acid but only a trace in selenomethionine. In pigs, selenomethionine has been shown to have higher availability than the form of Se-methyl-selenocysteine (Zhang et al., 2020). Therefore, after replacing fish meal with soybean meal, the essentiality of Se supplementation in the diet should be carefully reconsidered.

Poor growth performance was exhibited in giant grouper fed a soybean meal–based control diet compared with grouper fed an all–fish meal (reference) diet (Lin & Lin, 2021). The Se concentration in the control diet was 0.77 mg Se/kg, which was less than the Se concentration in the reference diet (1.07 mg Se/kg). In this study, Se supplementation in a soybean meal-based diet can linearly improve liver oxidative status and antioxidant enzyme capacity for giant grouper. However, Se supplementation in the diet did not present a positive effect on fish growth, suggesting that Se may not be the only factor that affects fish growth.

7.3 ANTIBIOTIC FREE OR REDUCED AQUAFEED

Pathogenic infection and disease outbreak occur more frequently in intensive culture systems, leading to retarded growth, inappetence, and death. In the past few decades, the use of antibiotics and chemotherapies has been considered to prevent or control disease in aquaculture. These chemicals are no longer recommended due to their spread of drug-resistant pathogens (superbugs), inhibition of the immune system, accumulation within aquatic animal tissues, and environmental pollution (Kuo et al., 2022). Thinking in terms of "antibiotic-free" or "reduced antibiotics" must be practiced in sustainable aquaculture. The immune system can be influenced by a wide range of factors, including diseases, pollutants, hormones, and diet (nutrition). The nutritional aspects of immune function are very important because diet can have a great impact on the immune response in fish. Amino acids, lipids and fatty acids, vitamins, and minerals are nutrients that have been studied for their effects on immune responses for aquatic animals.

In this chapter, we also discuss the reuse of agricultural waste and by-products. This objective is in line with the 12th item of the United Nations Sustainable Development Goals (SDGs): "Ensure sustainable consumption and production patterns."

7.3.1 ESSENTIAL NUTRIENTS

7.3.1.1 Amino Acids

Several studies report that amino acid supplementation in the diet enhances immune responses for fish and shrimp. For example, 0.5% tryptophan for sturgeon, 2.6–3.1% leucine and 2.7–3.6% arginine for golden pompano, 2.7% arginine for yellow catfish, 2% arginine and 1% glutamine for hybrid striped bass, and/or 1% arginine and 1% glutamine for a red drum diet have been reported to enhance immune responses of the fish (reviewed by Lin, 2016). On the basis of these reports, arginine and glutamine appear to be potential immunostimulants in diets for fish.

7.3.1.2 Lipids and Fatty Acids

Dietary lipid deficiency in the diet depressed nonspecific immune responses of Malabar grouper (*E. malabaricus*), and 4–12% of dietary lipid can maintain optimum immune responses for grouper (Lin & Shiau, 2003). Lin & Shiau (2007) indicated that Malabar grouper needs n-3 and n-6 fatty acids to maintain immunity. A ratio of n-3/n-6 fatty acids at 1–2 can improve grouper immune responses. Furthermore, enhanced immune responses were observed when the dietary docosahexaenoic acid (DHA)/eicosapentaenoic acid (EPA) ratio was greater than 1, indicating that DHA was superior to EPA in promoting grouper health (Wu et al., 2003).

7.3.1.3 Vitamins and Minerals

Vitamin C (L-ascorbic acid, AA) acts as a biological reducing agent for hydrogen transport. It is involved in many enzyme systems for hydroxylation of tryptophan, tyrosine, and praline and is also necessary for the formation of collagen and normal cartilage (NRC, 2011). Vitamin C is most widely investigated for its role in immunity for fish and shrimp. When comparing adequate levels for growth and immune responses, increasing the supplementation levels of vitamin C could effectively improve immune responses and resistance to pathogens (Lin & Tsai, 2020). For example, the optimal vitamin C requirement for the growth performance of Malabar grouper is a diet of 45 mg/kg, whereas that for immune response is a diet of 288 mg/kg (Lin & Shiau, 2005a). This suggests that an increase of vitamin C of 6.4 times can improve grouper immunity. For other species, similar results have been demonstrated in channel catfish, rainbow trout, turbot, gilthead sea bream, bagrid catfish, mrigal, Japanese sea bass, grouper, yellow croaker, Indian major carp, Japanese eel, and tiger shrimp (reviewed by Lin & Tsai, 2020). In general, in these studies, vitamin C has been shown to improve fish immunity 2 to 100 times.

Vitamin E functions as a lipid-soluble antioxidant, protecting biological membranes, lipoproteins, and lipid stores against oxidation. The antioxidative functions of vitamin E include scavenging of free radicals to terminate lipid peroxidation, which can initiate damage to unstable intracellular components, including membranes, nucleic acids, and enzymes and thereby result in pathological conditions. The estimated vitamin E requirements based on growth and immune responses for several fish and shrimp have been reviewed by Lin & Tsai (2020). This showed that a higher level of supplementation would improve the immunity of aquatic animals. Weight gain and liver oxidative status analyzed by broken line regression indicated that the optimal vitamin E diet for the juvenile grouper was a diet of 61–68 and 104–115 mg/kg in the 4 and 9% lipid diets, respectively (Lin & Shiau, 2005). The supplemented level of vitamin E that resulted in maximal nonspecific immune responses, including white blood cell count, O_{2^-} production ratio, lysozyme activity, and alternative complement activity was about 200–400 mg/kg. This suggests that 2–4 times vitamin E supplementation in the diet can enhance immune responses of the species. In addition, increased vitamin E supplementation in the diet also improves immune responses of fish, indicating 1.4–20 times vitamin E supplementation is recommended (Lin & Tsai, 2020).

Vitamin A is crucial in several physiological processes necessary for optimal animal function, including cell differentiation, reproduction, embryo development, development of epithelial cells from stem cells, and proper differentiation of immune cells. Rainbow trout fed a vitamin A-supplied diet resulted in higher total serum antiprotease, classical complement activity, and leukocyte migration but not serum immunoglobulin (Ig) level, lysozyme activity, and respiratory burst activity of phagocytes compared to fish fed a vitamin A-free diet (Thompson et al., 1995). Hernandez et al. (2007) observed that serum antibacterial activity was significantly higher in Japanese flounder fed diets containing 10,000 and 25,000 IU vitamin A/kg than in those fish fed the vitamin A-free diet.

Vitamin D functions to maintain calcium homeostasis together with two peptide hormones, calcitonin and parathyroid hormone. It is critically important for the development, growth, and maintenance of a healthy skeleton from birth until death. In addition to these functions, vitamin D has been found to play important roles in bone, skin, and blood cells; the secretion of insulin and prolactin; muscle function; immune and stress responses; and melanin synthesis. Cerezuela et al. (2008) conducted a study to assess the *in vivo* effect of cholecalciferol (vitamin D_3) on some innate immune parameters of the gilthead seabream (*Sparus aurata*). The tested humoral innate immune parameters (natural hemolytic complement activity and serum peroxidase level) were not increased by vitamin D_3 administration but, in contrast, decreased. Immunostimulant effects were found in the cellular innate immune parameters, especially cytotoxic and phagocytic activities. This agrees with the findings in higher vertebrates that demonstrated that vitamin D_3 not only promotes the induction of monocyte differentiation to macrophages but also modulates macrophage responses (Helming et al., 2005). The immunostimulant effect was greater in gilthead seabream on cellular immune parameters compared to humoral immune parameters, suggesting that receptors similar to those present in mammals are involved in the action of this vitamin in the immune systems of fish.

Vitamin B_1 (thiamin) was the first vitamin to be recognized. In animal tissue, thiamin occurs predominantly in a diphosphate form known as thiamin pyrophosphate (TPP). TPP is an essential cofactor for a number of important enzymatic steps in energy production, including both decarboxylation and transketolase reactions. Feng et al. (2011) reported that the survival rate, leucocyte phagocytic activity, lectin potency, acid phosphatase activity, lysozyme activity, total iron binding capacity, and immunoglobulin M content of Jian carp injected with *Aeromonas hydrophila* were all improved with an increase in dietary vitamin B_1 levels up to a 0.8–1.1 mg/kg diet.

The vitamin B_6 is the generic descriptor of 2-methylpyridine derivatives that have biological activity of pyridoxine. Pyridoxine is the main form found in plant products, whereas pyridoxal and pyridoxamine are the main forms found in animal tissue. All three forms are readily converted in animal tissue to the coenzyme forms, pyridoxal phosphate and pyridoxamine phosphate. Feng et al. (2010) reported that Jian carp fed diets with 5 mg/kg vitamin B_6 had higher white blood cell count, hemagglutination titer, lysozyme activities, acid phosphatase, and total iron binding capacity than fish fed a diet without supplementation with vitamin B_6 (1.7 mg B_6/kg in the base diet). Rohu (*Labeo rohita*) fed diets containing pyridoxine had a higher count of erythrocytes, hemoglobin content, total serum protein, albumin, globulin, nitroblue tetrazolium, and lysozyme activity and lower cortisol, blood glucose, and survival after challenge with *Aeromonas hydrophila* compared to the pyridoxine-deficient group (Akhtar et al., 2010). Akhtar et al. (2012) reported that dietary B_6 supplementation at a level of 100 mg/kg diet improved immune responses and resistance to high temperature stress in rohu.

Folate is used as a generic descriptor for folic acid and related compounds that qualitatively exhibit the biological activity of folic acid. Folate-dependent reactions are found in the metabolism of certain amino acids and the biosynthesis of purines and pyrimidines, along with the nucleotides found in DNA and RNA. A dietary folic acid level of 0.8 mg/kg diet is required for maximizing the growth of grouper; this amount is also adequate for nonspecific immune responses, including superoxide anion production and plasma lysozyme activity of the fish (Lin et al., 2011).

Pantothenic acid is a component of Coenzyme A (CoA), acyl CoA synthetase, and acyl carrier protein. The coenzyme form of the vitamin is therefore responsible for acyl group transfer reactions. Coenzyme A is required in reactions in which the carbon skeletons of glucose, fatty acids, and amino acids enter the energy-yielding tricarboxylic acid cycle. The acyl carrier protein is required for the synthesis of fatty acids. Wen et al. (2010) indicated that the survival rate of Jian carp after *Aeromonas hydrophila* challenge was higher with increasing dietary pantothenic acid levels. The optimal levels of pantothenic acid in the diet for leukocyte phagocytic activity and serum IgM content were respectively 42.2 and 47.2 mg/kg.

Inositol may exist in one of seven optically inactive forms and in one pair of optically active isomers. Only one of these forms, *myo*-inositol, possesses biological activity. Inositol is a biologically active cyclohexitol and occurs as a structural component in biological membranes, as does phosphatidylinositol. Red blood cell count, white blood cell count, phagocytosis activity, hemagglutination level, lysozyme activity, and Jian carp anti-*A. hydrophila* antibody titer improved with an increase in *myo*-inositol levels from a diet of 232.7 mg/kg to a diet of 687.3 mg/kg (Jiang et al., 2010).

Until now, few studies have focused on the effects of mineral levels on immune responses for aquatic animals. The growth and immune response estimates were similar in copper (Cu) and zinc (Zn) for tiger shrimp (*Penaeus monodon*) (Lee & Shiau, 2002; Shiau & Jiang, 2006). For Malabar grouper, the fish needs a higher dose of Se (1.38–2.02 mg/kg diet) to improve respiratory burst and lysozyme activities and plasma total immunoglobulin concentration (Lin & Shiau, 2006).

7.3.2 OTHER IMMUNOSTIMULANTS

Alternatively, immunostimulants made from plant extracts or marine resources are considered a strategy instead of antibiotics and chemotherapies. Immunostimulants are substances that stimulate the immune system by inducing activation or enhancing the biological activity of its components. These components, such as polysaccharides, alkaloids, flavonoids, pigments, phenolics, terpenoids, steroids, and essential oils, were shown to promote health benefits in fish and shrimp culture. The application of natural immunostimulants is a great advance for disease management in aquaculture and the environment. Several plant extracts were found to have immunostimulant properties, including seaweed polysaccharides (sodium alginate), water hyacinth, banana peel, cacao pod husks, and lemon peel.

7.3.2.1 Seaweed Polysaccharides

Some substances obtained from marine sources, mainly polysaccharides, can increase protection against infectious diseases. Sodium alginate has been extensively studied and is considered an immunostimulant. Fingerling orange-spotted grouper fed sodium alginate-containing diets was observed to promote growth performance and increase survival rates against *Streptococcus* sp. and iridovirus (Yeh et al., 2008). The anti-stress and immune responses of orange-spotted grouper were upregulated after being fed with a sodium alginate-containing diet (Lee et al., 2017). Survival rates of Taiwan abalone (*Haliotis diversicolor supertexta*) against *Vibrio parahaemolyticus* and its immune responses were enhanced when abalone was fed diets containing sodium alginate (Cheng & Yu, 2013). Both the injected method and the dietary administration of sodium alginate could increase immunity and resistance to *V. alginolyticus* of Pacific white shrimp (Cheng et al., 2004, 2005). In another white shrimp study, hot water-boiled brown seaweed meal was also demonstrate to enhance non-specific immune responses and resistance to *V. alginolyticus* for the shrimp (Lin et al., 2017b).

7.3.2.2 Water Hyacinth

Water hyacinth (*Eichhornia crassipes*) originated in South America, and it was introduced to many countries and distributed throughout the world. Due to its strong reproductive capacity and rapid consumption of dissolved oxygen in water, it can deteriorate water quality and has invaded Africa, Asia, and North America. Chang et al. (2013) indicated that giant freshwater prawn (*Macrobrachium rosenbergii*) fed with hot water extract from water hyacinth leaves for 12 days showed increased humoral and cellular immune responses, including hematopoiesis, prophenoloxidase system, respiratory bursts, superoxide dismutase activity, glutathione peroxidase activity, transglutaminase activity, coagulation, phagocytic activity, and clearance efficiency against *Lactococcus garvieae*. The study demonstrated that the hot water extract of water hyacinth leaves can act as an immunostimulant for prawns through dietary administration to enhance the prawn's immune responses.

7.3.2.3 Banana Peel

Banana, *Musa acuminate*, is the fruit whose production is second largest after citrus, accounting for about 17% of the world's total fruit production, and it is grown in more than 130 countries along the tropics and subtropics. However, the by-product of bananas, waste peels, causes an environmental and disposal problem. The main component of dietary fiber in banana peel consists of soluble and insoluble fractions such as lignin, pectin, cellulose, and hemicellulose. Additionally, flavonoids, tannins, phlobatannins, alkaloids, glycosides, and terpenoids were found to be present in the peels, which have been reported to exert antibacterial, antihypertensive, antidiabetic, and anti-inflammatory effects. Extracts of banana peel have been shown to increase the immunity of *M. rosenbergii*. Through injection administration, hot water extract of banana peel exerts an immunostimulating potential to enhance antibacterial activity and antihypothermal stress, immune responses, and resistance to diseases of the prawn (Rattanavichai & Cheng, 2014). The banana peel hot water extract diet supplement in the short term (within 32 days) and long-term (within 120 days) trail shows that the immune parameters and resistance to *Lactococcus garvieae* increased; especially, antihypothermal stress and growth performance are promoted in the short-term trail (Rattanavichai & Cheng, 2015; Rattanavichai et al., 2015).

7.3.2.4 Cacao Pod Husks

Cacao (*Theobroma cacao* L.) is commercially grown in South America. Its fruit contains approximately 30–40 beans, which are used mainly for the production of chocolate and cacao derivatives.

Application of Functional Aquafeed in Sustainable Aquaculture

These products show a significant increase in the growth rate of consumption around the world. According to an International Cocoa Organization estimate, the world gross cacao production was 4.7 million tons for 2019 and 2020 (ICCO, 2020). However, the pod husks of cacao, which account for 52–76% of the wet weight of the cacao pod, are the main by-product of cacao in addition to cacao beans. In other words, a large number of cacao pod husks, which were considered agricultural waste in the cacao industry in the past, were generated every year. Through the processing program, the various functions and bioactivity developed from cacao pod husks are good strategies to add value in the cacao production chain. Lee et al. (2020) indicate that pectin from cacao pod husk acts as an efficient immunostimulant for white shrimp. The injection of fresh cacao pod husk extract resulted in the enhancement of the cellular signaling pathways, specifically in the upregulation of TLR1, TLR3, and STAT gene expressions. This upregulation further induced an increase in both innate cellular and humoral immune responses, following increased resistance to pathogen and tolerance under hypothermal stress (Lee et al., 2020). Kuo et al. (2022) demonstrated that pectin from the dry cacao pod husk mediates growth performance, immune resistance responses, and carbohydrate metabolism of white shrimp through dietary administration. Cacao pod husk pectin also acts as a prebiotic and is combined with the probiotic *Lactobacillus plantarum* to form a symbiotic that improves the immunocompetence and growth of white shrimp (Kuo et al., 2021). All evidence indicates that the extract of the cacao pod husk has the potential to serve as an immunostimulant.

7.3.2.5 Lemon Peel

Lemon (*Citrus lemon*) is one of the most abundant citrus crops in the world, with an annual production of approximately 19 million tons in 2018 (FAO, www.fao.org/). The process of producing lemon juice, the main lemon product, produces 20–30% juice, leaving 50–60% waste material, consisting primarily of lemon peel. Therefore, a large amount of lemon peel is generated around the world each year. The disposal of this waste is costly and environmentally burdensome. To reduce this waste, many lemon peel applications have been implemented, including livestock feed, fertilizer, pectin extraction, bioethanol, and essential oil extraction. Lemons, like some herbal medicines, possess several biological effects, including anti-inflammatory, antimicrobial, immunomodulatory, antioxidant, and hepatoprotective properties. The main biocomponents, such as fiber, citric acid, ascorbic acid, minerals, flavonoids, carotenoids, limonoids, and essential oils, could provide a wide variety of beneficial health effects. Lemon essential oil supplementation at 0.3–0.6 g/kg diet has been reported to improve the plasma lysozyme and hepatic superoxide dismutase activity of orange-spotted grouper (Zhuo & Lin, 2021). Although lemon peel is considered to have a positive influence on health, to our knowledge, few studies have evaluated the application of citrus peel in aquafeed.

The fiber content of lemon peel was approximately 15% (dry matter), which presents a potential negative factor as a feed additive, although the fermentation process may be useful to lower the fiber content. To reduce the fiber content in lemon peel, our recent study used a probiotic *Lactobacillus plantarum* to degrade the fiber content in lemon peel from 19.60% to 13.55% during 24-hour fermentation processing (Zhuo et al., 2021a). Fermented lemon peel was used to evaluate its usefulness as an additive in fish feed. The first study demonstrated that 1–3% fermented lemon peel included in the diet can significantly improve enteritis induced by the dietary inclusion of soybean meal for Asian sea bass (Zhuo et al., 2021a). However, 5% supplementation with fermented lemon peel may lead to low immune responses and high oxidative stress for fish. In another grouper study, it was found that diets supplemented with 1–3% fermented lemon peel can significantly improve plasma lysozyme activity and resistance to the pathogen *Photobacterium damselae* for orange-spotted grouper (Zhuo et al., 2021b). Both studies clearly demonstrate that lemon peel has the potential to serve as a functional feed additive in aquafeed.

7.4 CONCLUSIONS AND PERSPECTIVES

The sustainable development of aquaculture is the most important issue for the continuous supply of human food in the future. The 2nd item, "Zero hunger"; the 12th item, "Responsible consumption and production"; and the 14th item, "Life below water," of the United Nations Sustainable Development Goals highlight the importance of aquaculture in the world. The 14th item especially indicates that human beings must protect and reduce their demands on marine resources. Reducing the dependence on fish meal in aquafeeds becomes an important issue in sustainable aquaculture. Although many alternative sources of protein for fish meal have been identified, previous research often overlooked the effects of plant-based protein on fish health status. This may lead to negative impacts, such as suppression of growth and immunity; poor feed utilization; low nutrient digestibility; and induction of oxidative stress, enteritis, and fatty liver in fish and shrimp fed plant protein–based diets. Several nutrients, including taurine, cholesterol, nucleotides, selenium, and organic acids, have been shown to improve these negative impacts. This also clearly indicates that these nutrients may be semi- (or conditionally) essential in a specific situation, such as fish fed a diet containing high levels of plant protein.

Another important issue in sustainable aquaculture is to reduce the use of antibiotics. In addition to the impacts of antibiotics on superbugs or environmental pollution, the use of antibiotics also raises food safety concerns for humans. The nutritional strategy has the potential to greatly aid aquaculture production through disease prevention and/or improvement of disease resistance. In past research, nutrients in diets were mainly evaluated for their benefits in growth and feed utilization but were overlooked for their potential in health control. In this chapter, we review the function of some nutrients and feed additives from agricultural waste/byproducts in improving immune responses and disease resistance for fish and shrimp. This information can be used to optimize feed formulation for the aquafeed industry. Thus, the development of functional feed for health control is expected.

When looking into available relevant papers, the issue of environment protection is paid less attention in the aquaculture industry compared with the topics we discussed previously. Nitrogen (N) and phosphorus (P) wastes and antibiotics or other chemicals in effluent from aqua farms may cause negative environmental impacts. Although a few countries have legislation to control pollutants in discharged water, the aquaculture industry in most countries lacks the concept of managing water pollutants. Functional feed for health control could reduce the use of antibiotics or other chemicals. Future work regarding of improvement of nutrient utilization and reduction of N or P excretion is necessary.

REFERENCES

Akhtar, M.S., Pal, A.K., Sahu, N.P., Alexander, C., Gupta, S.K., Choudhary, A.K., Jha, A.K., Rajan, M.G., 2010. Stress mitigating and immunomodulatory effect of dietary pyridoxine in *Labeo rohita* (Hamilton) fingerlings. Aquac. Res. 41, 991–1002.

Akhtar, M.S., Pal, A.K., Sahu, N.P., Alexander, C., Kumar, N., 2012. Effects of dietary pyridoxine on haemato-immunological responses of *Labeo rohita* fingerlings reared at higher water temperature. J. Anim. Physiol. Anim. Nutr. (Berl) 96, 581–590.

Cerezuela, R., Cuesta, A., Meseguer, J., Esteban, M.A., 2008. Effects of inulin on gilthead seabream (*Sparus aurata* L.) innate immune parameters. Fish Shellfish Immunol. 24, 663–668.

Chang, C.C., Tan, H.C., Cheng, W., 2013. Effects of dietary administration of water hyacinth (*Eichhornia crassipes*) extracts on the immune responses and disease resistance of giant freshwater prawn, *Macrobrachium rosenbergii*. Fish Shellfish Immunol. 35, 92–100.

Cheng, W., Liu, C.H., Kuo, C.M., Chen, J.C., 2005. Dietary administration of sodium alginate enhances the immune ability of white shrimp *Litopenaeus vannamei* and its resistance against *Vibrio alginolyticus*. Fish Shellfish Immunol. 18, 1–12.

Cheng, W., Liu, C.H., Yeh, S.T., Chen, J.C., 2004. The immune stimulatory effect of sodium alginate on the white shrimp *Litopenaeus vannamei* and its resistance against *Vibrio alginolyticus*. Fish Shellfish Immunol. 17, 41–51.

Cheng, W., Yu, J.S., 2013. Effects of the dietary administration of sodium alginate on the immune responses and disease resistance of Taiwan abalone, *Haliotis diversicolor* supertexta. Fish Shellfish Immunol. 34, 902–908.

Feng, L., He, W., Jiang, J., Liu, Y., Zhou, X.Q., 2010. Effects of dietary pyridoxine on disease resistance, immune responses and intestinal microflora in juvenile Jian carp (*Cyprinus carpio* var. Jian). Aquac. Nutr. 16, 254–261.

Feng, L., Huang, H.H., Liu, Y., Jiang, J., Jiang, W.D., Hu, K., Li, S.H., Zhou, X.Q., 2011. Effect of dietary thiamin supplement on immune responses and intestinal microflora in juvenile Jian carp (*Cyprinus carpio* var. Jian). Aquac. Nutr. 17, 557–569.

Helming, L., Böse, J., Ehrchen, J., Schiebe, S., Frahm, T., Geffers, R., Probst-Kepper, M., Balling, M.R., Lengeling, A., 2005.1alpha, 25-Dihydroxyvitamin D_3 is a potent suppressor of interferon gamma-mediated macrophage activation. Blood 106, 4351–4358.

Hernandez, L.H.H., Teshima, S., Koshio, S., Ishikawa, M., Tanaka, Y., Alam, S., 2007. Effects of vitamin A on growth, serum anti-bacterial activity and transaminase activities in the juvenile Japanese flounder, *Paralichthys olivaceus*. Aquaculture 262, 444–450.

ICCO, 2018. May 2020 quarterly bulletin. www.icco.org/about-us/icco-news/419-may-2020-quarterly-bulletin-of-cocoa-statistics.html.

Jiang, W.D., Feng, L., Liu, Y., Jiang, J., Hu, K., Li, S.H., Zhou, X.Q., 2010. Effects of graded levels of dietary *myo*-inositol on non-specific immune and specific immune parameters in juvenile Jian carp (*Cyprinus carpio* var. Jian). Aquac. Res. 41, 1413–1420.

Kuo, H.W., Chang, C.C., Cheng, W., 2021. Synbiotic combination of prebiotic, cacao pod husk pectin and probiotic, *Lactobacillus plantarum*, improve the immunocompetence and growth of *Litopenaeus vannamei*. Fish Shellfish Immunol. 118, 333–342.

Kuo, H.W., Chang, C.C., Cheng, W., 2022. Pectin from dry cacao pod husk mediates growth performance, immune resistance responses and carbohydrate metabolism of *Litopenaeus vannamei* through dietary administration. Aquaculture 548, 737613.

Lee, C.L., Kuo, H.W., Chang, C.C., Cheng, W., 2020. Injection of an extract of fresh cacao pod husks into *Litopenaeus vannamei* upregulates immune responses via innate immune signaling pathways. Fish Shellfish Immunol. 104, 545–556.

Lee, M.H., Shiau, S.Y., 2002. Dietary copper requirement of juvenile grass shrimp, *Penaeus monodon*, and effects on non-specific immune responses. Fish Shellfish Immunol. 13, 259–270.

Lee, P.P., Lin, Y.H., Chen, M.C., Cheng, W., 2017. Dietary administration of sodium alginate ameliorated stress and promoted immune resistance of grouper *Epinephelus coioides* under cold stress. Fish Shellfish Immunol. 65, 127–135.

Lim, C., Lückstädt, C., Webster, C.D., Kesius, P., 2015. Organic acids and their salts. In: Dietary Nutrients, Additives, and Fish Health, Lee, C.S., Lim, C., Gatlin, D.M. and Webster, C. (Eds.). Willey-Blackwell, Hoboken, NJ, pp. 305–320.

Lin, M.F., Shiau, S.Y., 2005a. Dietary L-ascorbic acid affects growth, nonspecific immune responses and disease resistance in juvenile grouper, *Epinephelus malabaricus*. Aquaculture 244, 215–221.

Lin, Y.H., 2016. Resolving disease issues with health supplements and management. 2016 DSM Aquaculture Conference Asia Pacific, Bangkok, Thailand, November 17.

Lin, Y.H., 2019. Regulation on lipid utilization and metabolism of fish fed high soybean meal-based diet. 12th Symposium of World's Chinese Scientists on Nutrition and Feeding of Finfish and Shellfish, Zhengzhou, PRC, October 14–18.

Lin, Y.H., Chen, Y.T., 2022. *Lactobacillus* spp. fermented soybean meal partially substitution to fish meal enhances innate immune responses and nutrient digestibility of white shrimp (*Litopenaeus vannamei*) fed diet with low fish meal. Aquaculture 548, 737634.

Lin, Y.H., Cheng, M.Y., 2017. Effects of dietary organic acid supplementation on the growth, nutrient digestibility and intestinal histology of the giant grouper *Epinephelus lanceolatus* fed a diet with soybean meal. Aquaculture 469, 106–111.

Lin, Y.H., Lin, H.Y., Shiau, S.Y., 2011. Dietary folic acid requirement of grouper, *Epinephelus malabaricus*, and its effects on non-specific immune responses. Aquaculture 317, 133–137.

Lin, Y.H., Lin, T.H., 2021. Comparison of selenomethionine and hydroxyselenomethionine on growth, selenium accumulation, and antioxidative capacity of giant grouper, *Epinephelus lanceolatus*, fed diet with soybean meal. Aquac. Nutr. 27, 2567–2574.

Lin, Y.H., Lu, R.M., 2020. Dietary taurine supplementation enhances growth and nutrient digestibility in giant grouper *Epinephelus lanceolatus* fed a diet with soybean meal. Aquac. Rep. 18, 100464.

Lin, Y.H., Mui, J.J., 2017. Comparison of dietary inclusion of commercial and fermented soybean meal on oxidative status and non-specific immune responses in white shrimp, *Litopenaeus vannamei*. Fish Shellfish Immunol. 63, 208–212.

Lin, Y.H., Mui, J.J., Lee, Y.C., 2017a. Effects of dietary soybean meal substituting to fish meal on growth, cholesterol status and molting signal gene expression in white shrimp, *Litopenaeus vannamei*. J. Fish. Soc. Taiwan 44, 1–9.

Lin, Y.H., Shiau, S.Y., 2003. Dietary lipid requirement of grouper, *Epinephelus malabaricus*, and effects on immune responses. Aquaculture 225, 243–250.

Lin, Y.H., Shiau, S.Y., 2005. Dietary vitamin E requirement of grouper, *Epinephelus malabaricus*, at two lipid levels, and their effects on immune responses. Aquaculture 248, 235–244.

Lin, Y.H., Shiau, S.Y., 2006. Effects of dietary selenium on non-specific immune responses in grouper, *Epinephelus malabaricus*. J. Fish. Soc. Taiwan 33, 315–323.

Lin, Y.H., Shiau, S.Y., 2007. Effects of dietary blend fish oil with corn oil in diet on the growth and non-specific immune responses of grouper, *Epinephelus malabaricus*. Aquac. Nutr. 13, 137–144.

Lin, Y.H., Su, Y.C., Cheng, W., 2017b. Simple heat processing of brown seaweed *Sargassum cristaefolium* supplementation in diet can improve growth, immune responses and survival to *Vibrio alginolyticus* of white shrimp, *Litopenaeus vannamei*. J. Mar. Sci. Tech. 25, 242–248.

Lin, Y.H., Tsai, T.T., 2020. Role of nutrition and immunity in fish: Cases in vitamins. J. Fish. Soc. Taiwan 47, 63–72.

National Research Council (NRC), 2011. Nutrient Requirements of Fish and Shrimp. National Academic Press, Washington, DC.

Rattanavichai, W., Chen, Y.N., Chang, C.C., Cheng, W., 2015. The effect of banana (*Musa acuminata*) peels hot-water extract on the immunity and resistance of giant freshwater prawn, *Macrobrachium rosenbergii* via dietary administration for a long term: Activity and gene transcription. Fish Shellfish Immunol. 46, 378–386.

Rattanavichai, W., Cheng, W., 2014. Effects of hot-water extract of banana (*Musa acuminata*) fruit's peel on the antibacterial activity, and anti-hypothermal stress, immune responses and disease resistance of the giant freshwater prawn, *Macrobrachium rosenbergii*. Fish Shellfish Immunol. 39, 326–335.

Rattanavichai, W., Cheng, W., 2015. Dietary supplement of banana (*Musa acuminata*) peels hot-water extract to enhance the growth, anti-hypothermal stress, immunity and disease resistance of the giant freshwater prawn, *Macrobrachium rosenbergii*. Fish Shellfish Immunol. 43, 415–426.

Salze, G.P., Davis, D.A., 2015. Taurine: A critical nutrient for future fish feeds. Aquaculture 437, 215–229.

Shiau, S.Y., Gabaudan, J., Lin, Y.H., 2015. Dietary nucleotide supplementation enhances immune responses and survival to *Streptococcus iniae* in hybrid tilapia fed diet containing low fish meal. Aquac. Rep. 2, 77–81.

Shiau, S.Y., Jiang, L.C., 2006. Dietary zinc requirements of grass shrimp, *Penaeus monodon*, and effects on immune responses. Aquaculture 254, 476–482.

Shiau, S.Y., Lin, Y.H., 2015. Dietary nucleotides supplementation enhances growth, immune responses and survival to *Vibrio alginolyticus* of juvenile white shrimp, *Litopenaeus vannamei*. J. Fish. Soc. Taiwan 42, 179–187.

Thompson, I., Choubert, G., Houlihan, D.F., Secombes, C.J., 1995. The effect of dietary vitamin A and astaxanthin on the immunocompetence of rainbow trout. Aquaculture 133, 91–102.

Wen, Z.P., Feng, L., Jiang, J., Liu, Y., Zhou, X.Q., 2010. Immune response, disease resistance and intestinal microflora of juvenile Jian carp (*Cyprinus carpio* var. Jian) fed graded levels of pantothenic acid. Aquac. Nutr. 16, 430–436.

Wu, F.C., Ting, Y.Y., Chen, H.Y., 2003. Dietary docosahexaenoic acid is more optimal than eicosapentaenoic acid affecting the level of cellular defence responses of the juvenile grouper *Epinephelus malabaricus*. Fish Shellfish Immunol. 14, 223–238.

Wu, T.M., Jiang, J.J., Lu, R.M., Lin, Y.H., 2020. Effects of dietary inclusion of soybean meal and cholesterol on the growth, cholesterol status, and metabolism of the giant grouper (*Epinephelus lanceolatus*). Aquac. Nutr. 26, 351–357.

Yeh, S.P., Chang, C.A., Chang, C.Y., Liu, C.H., Cheng, W., 2008. Dietary sodium alginate administration affects fingerling growth and resistance to *Streptococcus* sp. and iridovirus, and juvenile non-specific immune responses of the orange-spotted grouper, *Epinephelus coioides*. Fish Shellfish Immunol. 25, 19–27.

Yong, S.K.A., Abang Zamhari, D.N.J.B., Shapawi, R., Zhuo, L. C., Lin, Y.H., 2020. Physiological changes of giant grouper (*Epinephelus lanceolatus*) fed with high plant protein with and without supplementation of organic acid. Aquac. Rep. 18, 100499.

Zhang, K., Zhao, Q., Zhan, T., Han, Y., Tang, C., Zhang, J., 2020. Effect of different selenium sources on growth performance, tissue selenium content, meat quality, and selenoprotein gene expression in finishing pigs. Biol. Trace Element Res. 196, 463–471.

Zhuo, L.C., Abang Zamhari, D.N.J.B., Yong, A.S.K., Shapawi, R., Lin, Y.H., 2021a. Effects of fermented lemon peel supplementation in diet on growth, immune responses, and intestinal morphology of Asian sea bass, *Lates calcarifer*. Aquac. Rep. 21, 100801.

Zhuo, L.C., Chen, C.F., Lin, Y.H., 2021b. Dietary supplementation of fermented lemon peel enhances lysozyme activity and susceptibility to *Photobacterium damselae* for orange-spotted grouper, *Epinephelus coioides*. Fish Shellfish Immunol. 117, 248–252.

Zhuo, L.C., Lin, Y.H., 2021. Effects of dietary lemon essential oils on growth, biochemical parameters, nutrient digestibility, and histology of hepatocyte and intestinal villi of grouper, *Epinephelus coioides*. J. Fish. Soc. Taiwan 48, 1–10.

8 Sustainable Development for Shrimp Culture
A Critical Analysis

Kaushik Dey and Sanghamitra Sanyal

CONTENTS

8.1 Introduction 103
    8.1.1 Traditional Practices of Shrimp Farming 105
    8.1.2 The Extraordinary Expansion of Shrimp Culture: Fueled by Scientific Shrimp Farming Advancements 105
8.2 Sustainability Assessment of Shrimp Farming 106
    8.2.1 Sustainability of Supply and Quality of Inputs for Shrimp Culture 107
        8.2.1.1 Seed 107
        8.2.1.2 Feed 108
        8.2.1.3 Chemicals and Biological Products 110
    8.2.2 Environmental and Socio-Economic Sustainability 112
    8.2.3 Shrimp Disease Control for Sustainable Aquaculture 114
    8.2.4 The Use of Remote Sensing and Geographic Information System in Shrimp Farming 115
        8.2.4.1 Case Studies on Sustainable Shrimp Culture 116
8.3 Role of the Government in Maintaining Sustainable Shrimp Culture 117
8.4 Conclusion 119
    8.4.1 Acknowledgments 119
References 119

8.1 INTRODUCTION

The shrimp farming industry plays a significant role in the socio-economic uplift of coastal communities in the world. The continuous rise in the demand for shrimp worldwide has put ever-increasing pressure on producer countries to multiply their supply. This has resulted in the escalation of capture fishery activities and expanding the size of trawler fleets, as well as small vessels to facilitate over-exploitation of this very important marine resource. Most of the major shrimp fisheries at present are being harvested at full or near full capacity. A drop in the production trend would have a ruinous effect on the national economy. Given this, the last three decades have seen a multi-directional master plan for enhancing the production of shrimp, harnessing both capture and culture fishery with government and non-government entrepreneurial support.

    The terms "prawn" and "shrimp" have been used inconsistently. At the Prawn Symposium of the Indo-Pacific Fisheries Council held in Tokyo in 1955, it was decided that the term "prawn" should be applied to the Penaeids, Pandalids, and Palaemonids, while the use of "shrimp" should be restricted to the smaller forms belonging to other families. According to this definition, most of the forms of economic importance are prawns. Later, at the 1967 World Conference on the Biology and Culture of Shrimps and Prawns held in Mexico City, it was agreed that the term "prawn" will be

DOI: 10.1201/9781003326946-8

TABLE 8.1
Six Major Countries for Farmed Shrimp Production in 2018

Major Shrimp Production Countries in 2018

State	Estimated Production (t)
China	2,051,921
Indonesia	907,988
India	682,300
Ecuador	510,000
Vietnam	775,000
Thailand	362,910

Source: (FAO, 2020).

used for freshwater creatures only, and their marine/brackish water relatives will be called "shrimp" (Unnithan 2000).

A major share of the world's farm-raised shrimp is harvested in Asian countries, which offer a favorable site for shrimp aquaculture expansion, given the large brackish water areas convenient for farming, ideal climatic conditions, lower labor costs, and the Asian aquaculture tradition. During the 14years from 1984 to 1997, some 63 countries were listed in the Food and Agriculture Organization (FAO) Aquaculture statistics as having produced shrimp at one time or another. In 1997, nine of these countries didn't have any production announced. The potential shrimp-producing countries are Thailand, Indonesia, China, the Philippines, Taiwan, Vietnam, India, Ecuador, and Latin American countries (Table 8.1).

There are five different shrimp aquaculture practices, ranging from traditional to ultra-intensive techniques, but the most common techniques are extensive, semi-intensive, and intensive. These three categories are divided according to their stocking densities and the extent of management over grow-out parameters, that is, the level of inputs. The farmers who exercise extensive methods rely on cheap land and labor, naturally occurring seed stock and feeds, and a lack of regulations that allows the conversion of coastal lands to shrimp ponds. Little input is needed, so producers can relatively easy enter the industry. Semi-intensive and intensive farming practices require the aquaculturist to implement more control over the environment. Greater capital inputs, control of many grow-out parameters, and technical skills are required. The potential annual shrimp production (tons per hectare) from these systems is: 0.6–1.5 for extensive, 2–6 for semi-intensive, and 7–15 for intensive. The actual productivity is, however, much lower due to low-quality intake water, variable weather conditions, and especially disease problems. In 1999, the global average production was 650 kg shrimp per hectare pond, although most of the production was generated by semi-intensive practices. It must be accepted that the days of simple production-oriented shrimp farming are gone. Present-day production has to take note of not only the markets but a host of other technical issues as well as concerns about the environment. The subject matter of sustainable shrimp farming is broad, from farm-level management practices to integration of shrimp farming into coastal area management; shrimp health management; and policy, socio-economic, and legal issues.

The emergence of the SARS-CoV-2 coronavirus (COVID-19) pandemic has devitalized the shrimp farming industry, with the first lockdown unfortunately coinciding with the major annual shrimp farming cycle all over the world. The COVID-19 pandemic has had a notable impact on this sector. Fear of the virus and uncertainty about the future have impoverished many people working in this sector, both directly and indirectly. The lockdown all over the world formally started in mid-March 2020, closing all stages of shrimp culture all over the world. Furthermore, owing to stay-at-home orders, most hotels, restaurants, catering facilities, and local markets over the world were closed. Furthermore, the second wave of the pandemic has caused European countries to ban imports from other countries. Foreign buyers cancelling orders have had a significant impact on the shrimp farming industry. As a result, lower demand for shrimp products both in local and international markets has had a detrimental influence on income and livelihood patterns, particularly for small farmers. In general, they also lost money from farming because they were unable to market their harvests or were forced to sell at low prices.

8.1.1 Traditional Practices of Shrimp Farming

It has been regularly pointed out that countries that have significant success with aquaculture generally are those which have a long history and tradition associated with some form of aquaculture or related fishing activities. Shrimp has been cultured in South East Asia and China for centuries, utilizing traditional low-density methods. In Indonesia, the use of brackish water ponds, named *tambaks*, can be traced back as far as the 15th century. Commercial shrimp farming can be traced to the 1930s, when Japanese farmers spawned and cultured Kuruma shrimp (*Penaeus japonicus*) for the first time. By the 1960s, a tiny industry had developed in Japan. In India, the traditional knowledge of shrimp farming, popularly known as "trapping and holding," has been practiced frequently for decades in the low-lying brackish water confinements by Vembanad Lake in Kerala, known as *Pokkall fields*, and in the Sundarbans mangrove swamps of West Bengal, named *Bheries*. In the western hemisphere, shrimp were first cultured in Ecuador. This farming became a major profit-oriented activity in Ecuador by 1977. Shrimp farming spread from Ecuador to several other countries in Latin America, with the most significant production from Ecuador, Mexico, and Brazil. The major success with shrimp culture was in Taiwan through the development of an intensive culture method for *Penaeus monodon*. Shrimp farming in Vietnam started to develop from the beginning of the 1990s. Thailand has been one of the world's largest producers of farmed shrimp since 1991.

8.1.2 The Extraordinary Expansion of Shrimp Culture: Fueled by Scientific Shrimp Farming Advancements

China is the country with longest history of aquaculture in the world where shrimp culture relatively new industry. In 1981–1988, it experienced the first golden period of shrimp culture industry development after the full maturation of mass production hatchery techniques for *P. chinensis*. Peak production was maintained for five years, only ending in 1992 due to a serious outbreak of shrimp disease. The industry recovered slowly in next three years and regained its production mainly due to modifications in culture system and technique. Since 1998, China has been in a new era of growth in cultured shrimp production. The rapid growth was maintained due to the rapid expansion of culture of *P. vannamei*, an exotic shrimp species. Shrimp in Indonesia were a by-product of milkfish culture in traditional brackish water ponds. It became the main product of aquaculture ponds when appropriate culture technology, including mass-production seeds, became available in the 1980s.

In India, Menon (1954) notes that the traditional shrimp filtration system practice used a simple device for large-scale shrimp fishing that does not allow any time for shrimp to grow in fields. George et al. (1968) experimented to determine the viability of conventional filtration fields for introducing culture operations and discovered that these fields provide an active and acceptable biological environment for the survival and growth of shrimp. Muthu (1978) remarked on the scope for improving "culture practices" and production trends by way of propagating selective farming of shrimp at semi-intensive and intensive levels and evaluated the merits and demerits and ecological and techno-economic aspects of conventional practices. Based on this, the Central Marine Fisheries Research Institute (CMFRI), India, has been able to develop indigenous low-cost techniques for the culture of marine penaeid shrimp during its Fifth Five Year Plan (Silas 1978). Several Indian organizations, including the CMFRI, the National Institute of Oceanography, the Central Institute of Fisheries Education, the Konkan Krishi Vidyapeeth, some universities, and the All India Co-ordinated Research Project on "Brackish-Water Prawn and Fish Culture," have devoted significant resources to this area of study (Rao 1981). *P. indicus* and *P. mondon* are the most economically important shrimp species due to their rapid growth, enormous size, and great economic worth (Alagarswami 1981). Unnithan (1985) presented a package of practices of upgraded technology tested and proven at the Institute, which involves the exclusive stocking of seeds of commercially more important species of shrimp such as *P. indicus* and *P. mondon* proportionate to the area and productivity of the fields and growing them for definite periods to achieve good quality and maximum quantity of shrimp for greater profitability than the conventional shrimp filtration system. Apart from seasonal and perennial fields under traditional practice, other backwater and estuarine areas, including shallow brackish water canals in coconut groves and derelict water bodies in salt pan areas along the coastline, can be utilized for shrimp farming. The Central Marine Fisheries Research Institute has published detailed guidelines for the selection of sites and construction and maintenance of shrimp farms (Ramamurthy 1978; Unnithan 1985).

Ecuador stands out as an oddity among the major shrimp-producing countries. It is the only major shrimp-producing country outside Asia. As early as 1984, it was already producing more than the major shrimp-producing countries in Asia. Since it was totally dependent on wild fry, the industry remained at almost the same level until 1987 and was overtaken by China and Indonesia in 1985. Production picked up only when hatcheries started to be put up. Practically all the Latin American countries, from Mexico to Peru, have also started in the shrimp culture industry.

Vietnam has a long coastline of 3260 km, with a coastal area of over 1.5 million hectare and an exclusive economic zone of more than 1 million km^2. The natural features give Vietnam great potential for developing aquaculture using water bodies, coastal water bodies, and marine water areas. Shrimp farming in Vietnam started to develop from the beginning of the 1990s. Since 2001, white-leg shrimp farming in brackish water has become a major part of coastal economic development in Vietnam. More recently, new technologies have been adopted as a part of the process of moving towards more intensive farming methods.

Shrimp farming in Thailand has become a multi-billion industry and a major export earner. The rapid growth of shrimp culture in Thailand has led to an economic boom, especially in the coastal provinces of the eastern and southern regions. Shrimp farming can be characterized as a boom-and-bust industry, where the money earned in booms has not necessarily "trickled down" to the traditional coastal communities.

From the 1980s onwards, international and national organizations started to support poor and marginal rice farmers in adopting shrimp farming and pushed the commercial-scale shrimp culture industry to new heights but, unfortunately without addressing some of its ill effects in South-East Asian countries. While in the beginning, the shrimp industry was highly profitable, it was also found to be very risky, with viral diseases entering the country from Thailand in 1984 and destroying the whole industry at that time in South-East Asian countries. Some research institutes in Asian countries started to look into methods to reduce the risk of infection, including through crop rotation and water treatment, but the risk of disease remains to this day. As the industry matured, it started to address its negative impacts while drawing increasing criticism from both national and international groups. Since 2005, international and local non-governmental organization have raised questions about the sustainability of the sector, and different governmental groups and NGOs were formed to improve the quality of farmed shrimp, their traceability, and the social and environmental impacts of the sector.

8.2 SUSTAINABILITY ASSESSMENT OF SHRIMP FARMING

Sustainable development is development that meets the needs of the present without compromising the ability of future generations to meet their needs (Brundtl and Report 1987). According to Wurts (2000), sustainability is a systematic concept relating to the continuity of financial, social, institutional, and ecological aspects of human society. In a general sense, sustainability becomes an important requirement to survive in a world where there is a constant rise in the consumption of finite resources. He also states that developing nations' aquaculture industries must follow an autonomous framework to change production and market structures, recycle waste effluents, gather plankton, and stock densities.

Shrimp farming is often called the Blue Revolution. It has created huge employment opportunities in India, which is one of the strong points in favor of shrimp culture. With the advancement of scientific shrimp farming in India, not only did the yields increase, but it also attracted criticism on the grounds of generating adverse environmental and social impact, challenging the sustainability of the system. Shrimp culture is also characterized by inadequate and ill-implemented policies, conflicts of power and privilege, insufficient infrastructure, wage inequality, and so on. However, this industry has been plagued by discrepancies, controversies, and complaints from its inception. Thus, we need sustainability of shrimp culture based on natural resource management and conservation, with an emphasis on technological and institutional obligations to ensure a sustained supply of human requirements for current and future generations (Manoj and Vasudevan 2009).

8.2.1 Sustainability of Supply and Quality of Inputs for Shrimp Culture

8.2.1.1 Seed

The availability of quality seed is likely to be a major problem in the coming years, especially for the availability of brood stock. Efforts have to be made to develop a program of brood stock at the international level. Significant advances in domestication, selective breeding, and stock improvement in recent years have been made, especially in the Western Hemisphere, where several lines of domesticated shrimp are approaching the 36th generation in captivity. The domestication and selective breeding are slow in the Eastern Hemisphere, particularly in India, where no significant move has been made so far in this direction.

In most of the Asian shrimp farming countries, seed is still caught in the wild. The supply of seed cannot be guaranteed; its availability is generally seasonal, and it may carry disease. Approximately 80% of *Litopenaeus vannamei* farms in Asia are semi-intensively farmed. With survival rates of just 55%, Indian farms are significantly less productive than those in other countries, even in countries such as Thailand that have previously been ravaged by disease. Asia has thus far been spared a major disease outbreak, but its rapidly growing farms have low biosecurity standards, and therefore disease risk is high. Industry experts expect that Asian farmers will face a major disease outbreak eventually. Despite these challenges, Asia's shrimp producers are reluctant to change their methods: the industry is currently profitable, and the supply chain costs remain low. While it's true that Asia continues to dominate the global export market, growth rates are slowing, and inaction poses a high risk.

Reputational damage due to contaminated shrimp and unsustainable practices could affect profitability for years. Farms that do not comply with strict environmental standards, which are being more strongly enforced by local authorities, could be shut down. Post-larvae (PL) shrimp produced by hatcheries are critically important for farmers. High-quality PL production can improve grow-out farm survival rates as well as the quality and health of shrimp, ultimately benefiting the entire industry. Hence, hatcheries represent a crucial enabler.

Individual hatcheries should focus on improving quality by domesticating brood stock and implementing selective breeding practices to compete more effectively against the significant market power of integrated players. Because developing better PL involves genetic testing and investments in R&D, the implementation might be rather difficult for small hatcheries. Institutions and players with the necessary means should, therefore, continue to support small hatcheries in these efforts.

Due to their large size and high price, *P. monodon* (Figure 8.1) and *P. indicus* are generally considered for farming. It has also been seen that both these species are suitable for farming in the Asian environment. Apart from these candidate species, other commercially important species such as *Metapenaeus ensis*, *M. monoceros*, *M. brevicornis*, *P. semisulcatus*, and *P. merguiensis* are also potential species that can be grown in Asia. Another potential candidate species that is flooding the international market is the white leg shrimp, *Litopenaeus vannamei* (Figure 8.2).

Due to the sudden outbreak of COVID-19, hatcheries in various stages of production were forced to stop operating, and the stocks were terminated, anticipating poor demand. Major issues faced by

FIGURE 8.1 Post-larvae of *Penaeus monodon*.

FIGURE 8.2 Healthy *Litopenaeus vannamei*.

the hatchery sector included reduced demand for seed due to cancellations of seed bookings and postponement of pond stocking, and there were some issues in importing specific pathogen-free (SPF) broods, as the reservations made by hatcheries for the import of about 7600 Vannamei brood stock were canceled during the lockdown period, along with two Monodon broodstock shipments and one SPF PL shipment.

8.2.1.2 Feed

The development and use of compound feed have been major advancements in the successful expansion of shrimp farming. As in most animal-production enterprises, feed accounts for the largest operating cost (about 60–70%), and proper feed management is crucial for the profitability of shrimp farming. Feed management techniques (Table 8.2) are as important as feed quality in both

TABLE 8.2
Feed Management Techniques for Shrimp Based on Body Weight

Shrimp Live Body Weight (g)	Recommended Feeding Rate (% body weight/day)
2–3	8.0–7.0
3–5	7.0–5.5
5–10	5.5–4.5
10–15	4.5–3.8
15–20	3.8–3.2
20–25	3.2–2.9
25–30	2.9–2.5
30–35	2.5–2.3
35–40	2.3–2.1

TABLE 8.3
Recommended Nutrient Levels for Shrimp on Percentage Fed Basis

Shrimp Size (gm)	Protein (%)	Fat (%)	Fiber (%)	Ash (%)	Moisture (%)	Calcium (%)	Phosphorus (%)
0.0–0.5	45	7.5	0.4*	0.15*	0.12*	0.2.3*	1.5**
0.5–3.0	40	6.7	0.4*	0.15*	0.12*	0.2.3*	1.5**
3.0–15.0	38	6.3	0.4*	0.15*	0.12*	0.2.3*	1.5**
15.0–40.0	36	6	0.4*	0.15*	0.12*	0.2.3*	1.5**

*Maximum recommended. **Minimum recommended.

improved traditional and extensive shrimp farming systems. The best shrimp feed will be at best an expensive fertilizer if not managed properly.

The shrimp farming sector has received criticism in recent years for the excessive use of fishmeal in formulated feeds. While the use of fishmeal poses no present threat to the sustainability of marine fisheries, it is important to develop fishmeal substitutes over time. In Asia, where *P. monodon* is the predominant cultured shrimp species, feed protein levels have not been significantly reduced. Today, the protein levels average around 38%.

The use of good-quality feed (Table 8.3) will improve shrimp production and profits and minimize the environmental pollution from shrimp farming.

An ideal feed conversion ratio (FCR) always results in model growth rate, healthy shrimp, and clean pond bottom conditions. Only a superior quality of feed can achieve an FCR of 1.2. According to recent data, an FCR as low as 1.2 has been achieved, but many farmers are still obtaining FCRs higher than 2.2. Therefore, besides feed management, the FCR is also closely related to the quality of feed. Model-quality shrimp feed must be highly palatable. Since shrimp are a slow-feeder animal, the water stability of suitable feed should be over two hours for *P. monodon*. Feed quality will rapidly deteriorate if the feed is not packed well and properly stored. The feed should be stored in a dry, cool, and well-ventilated place to maintain consistent moisture and temperature. The feed should not be stored in direct sunlight and should not be kept longer than three months from the time of processing. Spoiled or old feed should not be used. The feed market in Asia is expected to grow at 11% per year through 2022, in line with Asia's overall shrimp market. While some large corporate farms buy feed directly from feed mills, the vast majority of farmers use a well-established dealer network. Water quality parameters required for maximum feed efficiency and maximum growth of shrimp are given in Table 8.4.

TABLE 8.4
Water Quality Parameters for Maximum Feed Efficiency and Maximum Growth of Shrimp Culture

Water Parameters	Optimum Level
Dissolved oxygen	3.5–4 ppm
Salinity	10–25 ppt
Water temperature	26–32°C
pH	6.8–8.7
Total nitrite nitrogen	1.0 ppm
Total ammonia (less than)	1.0 ppm
Biological oxygen demand (BOD)	10 ppm
Chemical oxygen demand (COD)	70 ppm
Transparency	35 cm
Carbon dioxide (less than)	10 ppm
Sulfide (less than)	0.003 ppm

During the pandemic period, the aquaculture sector has faced issues related to labor as migrant laborers returned to their native states/villages; feed-related issues as prices have gone up due to a lack of raw materials for feed manufacture and movement of feed also being restricted; fuel shortage—a shortage of diesel in generators for running pumps and aerators; shortage of seed due to lack of skilled labor and movement restrictions; and reduced intake by processing plants due to shortage in labor and uncertainty about the international market, leading to lower procurement from farmers by exporters. Panic harvest due to uncertainty and shortage of feed and diesel, farmers resorted to panic harvesting during the initial months of the 2020 and low product price for product. Due to panic harvesting, reduced demand from exporters, lack of adequate storage facilities, and uncertainty of markets, prices fell during the first half of 2020. The marketing of shrimp seed and feed should be regulated to ensure quality sustainability.

8.2.1.3 Chemicals and Biological Products

Chemicals and drugs used in aquaculture include those associated with structural material, soil and water treatment, antibacterial agents, therapeutants, pesticides, feed additives, anesthetics, immunostimulants, and hormones. Chemicals and drugs presently in use are mostly derived from agriculture or veterinary and have never been tested and evaluated specifically from the perspective of their effects on the aquatic environment. The use of many chemicals and drugs in aquaculture, if carried out properly, can be regarded as wholly beneficial, with no attendant adverse environmental effects or increased risks to the health of farm workers. However, the indiscriminate use of chemicals and drugs, especially those which are banned, may harm the environment and also incur penalties for the shrimp farming sector, including:

A. International trade difficulties arising from drug residues.
B. The potential for loss of efficacy of prophylactics/antibacterial agents.
C. Increased demand for and complexity of wastewater treatment.

The unscientific use of antibiotics, drugs, and other pharmacologically active compounds can hurt human health and also the environment. Most of the countries importing marine products from India do not permit any residue level of banned antibiotics and chemicals.

Sustainable Development for Shrimp Culture

Therefore, from the export angle, we must also ensure that shrimp farmers do not use antibiotics that have the potential to harm human health and adversely impact the environment.

The Marine Products Export Development Authority (MPEDA), a statutory body established in 1972 under the Ministry of Commerce & Industry in India, has prepared a list of drugs, chemicals, and other pharmacologically active compounds presently in use in shrimp farming. The MPEDA has also prepared a detailed monitoring plan for testing antibiotic and chemical residues in farmed shrimp. The monitoring plan envisages the collection of samples from shrimp farms and processing units directly by MPEDA and testing them for pesticide and antibiotic residues. In some Asian countries, several unauthorized "aqua shops" sell antibiotics, drugs and chemicals, and other medical formulations for shrimp culture, disregarding the ban on several of these products. Most of the products marketed are without labels and any description of their composition/usage. It was also noted that under the pretext of probiotics, there are several preparations marketed that contain antibiotics.

A list of 20 antibiotics, drugs, and pharmacologically active substances and four antibiotics for which a certain level of residue is permissible in the finished product was considered for shrimp culture by the Aquaculture Authority of India, as shown in Table 8.5. It was recommended that the body might consider including a provision in the approval given to shrimp farmers prohibiting the use of banned antibiotics and pharmacologically active substances during shrimp culture operations.

TABLE 8.5
List of Antibiotics and other Pharmacologically Active Substances Banned by Aquaculture Authority

Antibiotics and Other Pharmacologically Active Substances	Maximum Permissible Residual Level (in ppm)
Chloramphenicol	Nil
Nitrofurans, including furaltadone, furazolidone, fury 1 furamide, nifuratel, nifuroxime, nifurprazine, nitrofurantoin, nitrofurazone	Nil
Neomycin	Nil
Nalidixic acid	Nil
Sulphamethoxazole	Nil
Aristolochiaspp and preparations thereof	Nil
Chloroform	Nil
Chlorpromazine	Nil
Colchicine	Nil
Dapsone	Nil
Dimetridazole	Nil
Metronidazole	Nil
Ronidazole	Nil
Ipronidazole	Nil
Other nitroimidazoles	Nil
Clenbuterol	Nil
Diethyistilbestrol (des)	Nil
Sulfonamidedrugs (except approved sulfadimethoxine, sulfabromomethazine, and sulfaethoxy pyridazine)	Nil
Fluroquinolones	Nil
Glycopeptides	Nil

This will enable the aquaculture authority to regulate the use of banned antibiotics, chemicals, and so on and to cancel approval if the ban is violated by farmers. This ban would in no way hamper farming practices, as there are several safe alternative compounds available to be used by shrimp farmers during a crisis.

The awareness program was organized for achieving the objectives of regulating the use of antibiotics, drugs, chemicals, and so on in shrimp farming and to educate shrimp farmers and those associated with the manufacture, supply, and marketing of such chemicals. This program was implemented with the active cooperation of the local government body. Mobile and modern laboratories were set up to assist shrimp farmers in the management of animal health and water quality at farms and for testing antibiotic residues in farmed shrimp. The use of probiotics must be regulated, as they contain live organisms, and probiotics of foreign origin may have the potential to introduce exotic species, the impact of which may not be known. The Ministry of Health and Family Welfare, India, issued directives to drug manufacturers to label containers (bottles, packets, sachets, etc.) of veterinary-grade drugs with "Not for Use in Shrimp Culture." This would help check the rampant use of veterinary-grade drugs in shrimp farming.

8.2.2 Environmental and Socio-Economic Sustainability

A variety of socio-economic and environmental impacts have been described (see Figure 8.3) in different parts of the coastal area, including displacement of local people as a result of increased land value, decreased land quality (e.g., salinization, acidification, subsidence), resource appropriation (government allocation through concessions of previously held common property resources, which had been used for a variety of subsistence uses by local people), interference with navigational rights, and conflict with the shrimp industry.

It has additionally been suggested that the move from subsistence agriculture and the use of natural wealth to employment on massive shrimp farms could represent a reduction in quality of life. This is a political issue, not a scientific one, and it is the job of scientists and economists to contribute direction on these issues rather than making value judgments.

Environmental topics have always been a point of debate in shrimp culture industry development. While the harvest from capture fisheries around the world has stagnated, aquaculture is seen as a sound choice to upgrade shrimp production and plays an imperative part in giving nourishment and nutritional security. Be that as it may, the shrimp culture industry has been immovably restricted by

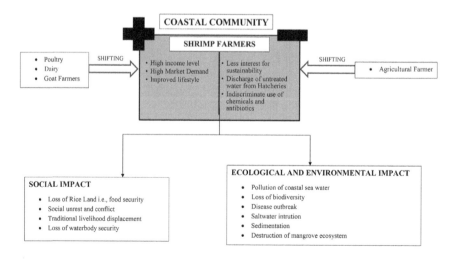

FIGURE 8.3 A variety of socio-economic and environmental impacts on coastal communities.

Sustainable Development for Shrimp Culture

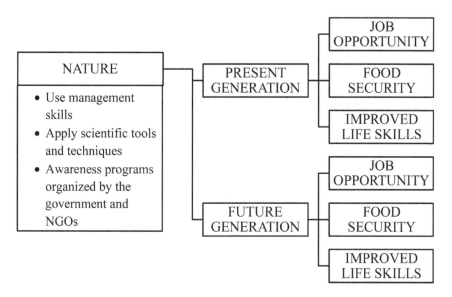

FIGURE 8.4 Basic characteristics of sustainable shrimp culture.

environmental groups. In Asia, on numerous events, legal interventions have been sought to curtail the shrimp industry to conserve the coastal environment.

The increasing demand, supply, and value of shrimp have led to unsustainable farming practices and environmental, social, and economic impacts. There's concurrence with nearby communities, NGOs, researchers, worldwide organizations, governments, and the shrimp industry that there's a demand for the industry to alter, to be more environmentally and socio-economically conscious, and, along with the customers and other major buyers, they are driving the key segments of the shrimp industry to take responsibility for their execution. Numerous related international, national, and local technical policy and legislative guidelines have been published by international organizations, the private sector, and NGOs, addressing the shrimp industry regarding environmental sustainability. These include codes of conduct, codes of practice, better management practices, and certification schemes, but many of them are voluntary, and there can be confusion about which to follow. Raising awareness of consumers, especially in coastal areas, about where shrimp come from and how they are produced may also help to drive the promotion of ecologically sustainable and socially respectable methods of farming shrimp. Some basic characteristics of sustainable shrimp culture are summarized in Figure 8.4.

The number of shrimp farms and areas under cultivation have increased considerably during the last two decades. In many areas, shrimp farms have developed close to one another along creeks and estuarine watercourses. The Kandaleru creek in Andhra Pradesh, India, is an excellent example of farms set up in large-scale clusters. As sustained development of shrimp culture relies on a source of good-quality water, overdevelopment of shrimp farms—either through management intensification or increased farm area—along a creek can impact estuarine water quality to levels unacceptable for shrimp farming. This fact brings into focus the urgent need for scientific investigations on carrying capacity, which will address not only the physical regions of shrimp farms but also their density and geographical distribution along a watercourse. The orderly development of shrimp farms will, therefore, need considerable attention from farmers, as well as planners and research establishments. Among the many good management practices that are currently in vogue and adopted by the farmers in the country, low stocking densities have proven successful in attaining sustainability.

The Aquaculture Authority permits the stocking of post larvae of to up to six per square meter (m^2) for farms within the coastal regulation zone (CRZ) and up to 10 m^2 outside the CRZ. It is presently claimed that a large number of farms in the country are embracing low stocking densities with great success. In terms of economics, the low stocking densities are also working well. Adoption of low stocking densities will be one of the key elements of sustainability in the years to come and needs to be promoted among shrimp farmers.

The introduction of social values to sustainability goals implies a much more complex and contentious debate, and those focused on ecology tend to strongly resist non-ecological interpretations. The most important factor is that the village community, which contributes labor, resources, and energy for aquaculture production, must benefit from the substantial profits obtained. Mitigation of social issues can most often be resolved by active participation and consultation by all stakeholders in the planning, development, and management of aquaculture. Conflict resolution and enabling environments at a local community level should exist for common resources such as seafront and freshwater resources to both aquaculture units and coastal communities.

8.2.3 Shrimp Disease Control for Sustainable Aquaculture

Shrimp aquaculture's viability and development are substantially jeopardized due to serious ecological and pathological problems that afflict the great majority of global shrimp farms. Prevention and control of diseases are now priorities for the durability of this industry. Field and mobile laboratories with adequate facilities are essential to provide a first-hand diagnosis in coastal areas. Elaborate and comprehensive guidelines should be prepared on sustainable shrimp farming and translated into vernacular languages so that farmers can be properly educated. This would help to mitigate problems, especially those related to disease and health management. Shrimp immunology is a key element in establishing strategies for the control of diseases in shrimp aquaculture. The CMFRI has been devoting much attention to investigations of the problems of disease among shrimp ever since serious pathological conditions were encountered in culture systems, including hatcheries, often leading to mass mortality of the stock. Pillai (1984) described the various types of diseases encountered in brackish water and marine environments, matters relating to defense against infection, prophylaxis and disease check-up, sampling techniques for disease diagnosis, classification of pathogens, screening of bacteria for identification, and methods to be adopted for the dispatch of diseased specimens for study. Rao (1983) made a comprehensive review of investigations on shrimp diseases in India and abroad, supplemented with his observations.

Various infectious diseases are caused by viruses (monodon baculovirus disease, hepatopancreatic parvo-like virus disease, yellow-head disease, white spot disease, infectious hepatopancreatic and lymphoid organ necrosis), bacteria (luminous vibriosis, vibriosis), fungi (larval mycosis), protozoa (black gill disease, surface fouling diseases, microsporidiosis/cotton shrimp disease/milk shrimp disease), trematodes, cestodes, nematodes, leaches, and parasitic crustacea; non-infectious diseases (soft-shell syndrome), including those caused by nutritional deficiencies, pollution (Figure 8.3), environmental stress, toxic algal blooms, and so on; and their remedial measures have been dealt with. The adoption of suitable strategies and practices aimed at a sustainable level of production alone can ensure healthy growth of the shrimp farming industry. Treatment cannot be carried out effectively when shrimp diseases occur in a pond. The best way to get rid of diseases is by practicing good farm management or prevention.

Infectious diseases in shrimp are one of the significant constraints in shrimp aquaculture across the globe. The alarming increase of antibiotic resistance in bacteria has rendered antibiotic therapy a controversial subject today. Therefore, the practice of antibiotic therapy in aquaculture needs to be dissuaded and appropriate alternatives need to be researched. For the past few decades, the fields of nanotechnology and biotechnology have proposed novel and effective solutions involving the use of nanoparticles, biofilm-based vaccines, algal extracts, phytobiotics, probiotics, prebiotics, and symbiotics, compounds of biological origin that are safe to use in aquaculture settings, to combat infectious diseases (Seethalakshmi et al. 2021). With the advent of green and bio-nanotechnology, nanomaterials are also becoming a safer alternative to antibiotic therapy. Vaccines developed from antigenic components of bacterial biofilms are more promising than regular vaccines synthesized from antigens of planktonic forms. Some of these methods have extended applications in shrimp aquaculture in the form of immunomodulants, diagnostic tools, and drug and vaccine carriers. The hazards of chemotherapy in shrimp aquaculture can be overcome by replacing antibiotics and other chemical agents with these new approaches. Adopting these strategies makes aquaculture-based food more organic and consumer-friendly, which helps in establishing sustainable aquaculture.

The introduction of modern diagnostic tools such as polymerase chain reaction (PCR) techniques to check the presence or absence of pathogens in the shrimp post-larvae before stocking has helped reduce risk. Other protocols relating to pond chlorination, pond treatment and cleanliness, and bio-security are additional developments that considerably reduce the chances of disease spread. In addition, new management techniques such as bio-remediation through various microorganisms, enzymes, and probiotics added to feed are also proving useful, and their use is advocated.

The sustainability of aquaculture will depend on the selection of disease-resistant shrimp, that is, to develop research in immunology and genetics at the same time. The development of strategies for prophylaxis and control of shrimp diseases could be aided by the establishment of a collaborative network to contribute to progress in basic knowledge of penaeid immunity.

Before harvesting and/or exporting, shrimp should be examined for their health, hygienic quality, and safety for consumers. Unhealthy shrimp, which are easily recognized through their appearance, will not be acceptable to consumers, and market value could be reduced. Unhealthy shrimp should be treated before harvesting or removed during harvesting and processing if the proportion of unhealthy shrimp in the stock is low.

Human pathogenic organisms could contaminate shrimp during harvesting, storage, and processing. Therefore, samples of shrimp should be sent to a reliable laboratory to conduct the necessary tests to certify the hygienic quality of the products before exporting or sending them to market. The harvested shrimp should also be checked for antibiotics and heavy metal residues before export.

8.2.4 The Use of Remote Sensing and Geographic Information System in Shrimp Farming

Remote sensing along with Geographic Information System (GIS) can play a major role in sustainable shrimp culture development by providing information on land use/land cover, water quality, productivity, tidal influence, and coastal infrastructure. These tools help to maintain the sustainability of shrimp culture through proper site selection by considering the impact of the development on other land-use activities like agriculture, protected areas like sanctuaries, and human use that are part of the same ecosystem. These tools also provide an image of the same location at periodic intervals. This helps us to keep a record of the images with changing environments for evaluation.

The term "remote sensing" is broadly defined as technique(s) for collecting images or other data about an object from measurements made at a distance from the object and can refer, for instance, to satellite imagery, aerial photographs, or ocean bathymetry explored from a ship using echo sounder data. However, in the present context, only optical images acquired by spaceborne or airborne sensors are considered. Over the last few decades, remote sensing technology has been used increasingly by the scientific community to describe and monitor a variety of systems on a local or global scale. This technology has evolved from pure visual imagery (e.g., panchromatic aerial photographs) to multispectral imagery (e.g., thematic mapper). Orbital remote sensing data, due to their synoptic and multispectral nature, provide a wide range of information over inaccessible and large areas at frequent intervals. This has made remote sensing technology a useful tool in coastal resource assessment (Rajitha et al. 2007).

A large number of earth observation satellites have orbited and are orbiting our planet to provide frequent imagery of its surface. From these satellites, many can potentially provide useful information for the sustainable management of shrimp culture. The great advantage of satellite imagery is the possibility of multitemporal evaluation and low costs compared to prolonged field sampling. The concentrations of optically active water constituents, namely chlorophyll, colored dissolved organic matter (CDOM), and total suspended solids (TSSs), can be estimated from satellite images by the interpretation of the received radiance at the sensor at different wavelengths (Gordon and Morel 1983).

GIS is widely used as a tool to digitize remotely sensed or cartographic data complemented by various ground-truth data, which are geo-coded using a global positioning system (GPS). GIS can be used to analyze the spatial characteristics of the data over various digital layers. GIS can be described as databases where the information is spatially referenced; what has made GIS so popular is the fact that the spatial referencing of information is related to maps. It is the manipulation and analysis of the spatial database and the display of maps with relative speed and ease that is the trademark of GIS. GIS has proved useful in assessing impacts on aquatic resources and environments for development projects involving land and water use, in aquaculture site selection concerning ecological and socio-economic variables, in space and resource allocation to conflicting types of use, and in aquaculture development planning and environmental impact monitoring.

8.2.4.1 Case Studies on Sustainable Shrimp Culture

The case studies in this section disclose the achievements possible by integrating GIS and remote sensing for coastal management with special attention to sustainable shrimp aquaculture.

In India, site selection for shrimp farming using GIS was achieved by Gupta (1995) and Ramesh and Rajkumar (1996). Vadlapudi (1996) carried out a study on the identification and quantification of changes in a mangrove forest using remote sensing and GIS in the Godavari delta. Farmers of the East Godavari district of Andhra Pradesh, India, have converted some portions of their paddy fields into shrimp farms, as observed by Kumaran et al. (2003).

In the last decade, following an upsurge in shrimp farming activity, about 13 km^2 of the mangrove area near Kakinada in the East Godavari district was removed (Satyanarayan et al. 2002).

Hein (2000) conducted a study on the impact of shrimp farming on mangroves along the east coast of India. An IRS LISS I image from1990 and LISS III image from1999 were used to identify the impact of shrimp farming in the Godavari delta. The results of the study revealed 80% of conversions of mangroves into shrimp ponds in the study area. Salam and Ross (2000) conducted a study on optimizing site selection for the development of shrimp and mud crab culture in southwestern Bangladesh using remote sensing and GIS. Arnold et al. (2000) carried out a study on a GIS-based approach for identifying appropriate sites for hard clam culture in Florida.

Hossain et al. (2003) carried out a study on monitoring shrimp farming development from space: a remote sensing and GIS approach in the Kandleru Creek area, Andhra Pradesh, India. The study encloses an area of 256.64 km^2 that covers the major part of the Kandleru Creek area where drastic alteration of land use or land cover can be found, along with mushrooming of shrimp farms. The results of this study disclosed an unrestricted and unintended growth of shrimp farms in the study area. It may have helped to boost the economy of rural people, but the environmental complications, which are already apparent, must be addressed for sustainable growth of the shrimp industry.

For identifying potential brackish water areas through remote sensing and GIS, Jayanthi and Rekha (2004) conducted a study in the Krishna district of Andhra Pradesh, India. According to this study, remote sensing along with GIS could be an intrinsic part of successful management because it has the facility for regular updating with time series information based on a well-defined user interface and appropriate provision for varying nature of data in the system design. Tong et al. (2004) conducted a study on assessment of the impact of shrimp aquaculture in the Mekong delta in Vietnam. They used spot scenes for identification of mangrove types and for delineation of landscape units.

Karthik et al. (2005) carried out a study on brackish water site selection in Palghar taluk of Thane district of Maharashtra, India, using remote sensing and GIS. This study demonstrated the potential use of remote sensing, GIS, and GPS for aquaculture site selection and planning. The weighted overlay method was adopted in GIS to delineate the potential area into four major categories as highly suitable, suitable, moderately suitable, and unsuitable. The study revealed that out of the total study area, 0.377% is highly suitable, 9.873% is suitable, 1.772% is moderately suitable, 85.02% is unsuitable, and 2.951% is already under aqua farms. Giap et al. (2005) conducted a study on land evaluation for shrimp farming in Haiphong in Vietnam. This study was conducted to identify appropriate sites for shrimp farming development in the study area using geographical information systems.

Computational intelligence (CI) techniques such as artificial neural networks (ANNs) provide the capability to better examine complex data without requiring detailed knowledge about the underlying physical system. ANN techniques have been used to accurately estimate bio-optical parameters in complex coastal aquatic environments from remotely sensed data by employing special features such as the ability to learn from input data, adaptive behavior, handling of non-linear systems, flexibility in choice of inputs, and resilience against noise. ANNs have proved successful in retrieving water quality information from space platforms. The parallel computing structure arises from neurons being interconnected by a network; ANNs probably hold more potential than regression models for retrieving water quality parameters from satellite images.

8.3 ROLE OF THE GOVERNMENT IN MAINTAINING SUSTAINABLE SHRIMP CULTURE

Some initiatives have been taken by the governments of shrimp-producing countries for maintaining sustainable shrimp culture. Governments and shrimp industrialists must ensure that shrimp culture practices become systematic, eco-friendly, and socially responsible.

Since the late 1970s, the Chinese government has been giving high importance to the establishment of the legal system and regulations to regulate various aquaculture-related activities. A "Technical Code for Shrimp Culture to Produce Healthy Food" was issued by the Ministry of Agriculture in 2001. The Chinese government is pushing very hard to establish and implement various laws and regulations on various social activities and relationships.

For the first three decades, the Thai government supported the promotion of shrimp culture, mainly through two agencies, the Department of Fisheries (DoF) and the Board of Investment (BoI). Since the 1970s, the DoF has undertaken extension services to promote shrimp farming; the BoI has promoted investment projects through support to industry and interim measures during falling shrimp prices or rising energy costs. In response to the loss of mangrove forests, the Thai government implemented conservation policies. Recently, the Department of Pollution Control (DPC) under the Minister of Science, Technology, and Environment (MOSTE) commissioned the Network for Aquaculture in Asia-Pacific (NACA) to undertake a study on water pollution from coastal fisheries (NACA 1969). The Office of Agricultural Economics (OAE) under the Ministry of Agriculture and Agricultural Co-operatives (MOAC) commissioned the Mekong International Development Council (MIDAS) to make recommendations on coastal resource management strategies (MIDAS Agronomic 1995).

The government of Vietnam allows farmers to convert low productive and saline rice fields, uncultivated areas, and salt pans in coastal areas into aquaculture ponds.

The government of Indonesia has since extended the following assistance to shrimp farmers:

A. construction/rehabilitation of migration canals in brackish water pond areas
B. disease prevention in grow-out ponds
C. seed production technology
D. advice on site selection, pond arrangement, and post-harvest through extension

Indian government research institutes like the Central Institute of Brackish Water Aquaculture (CIBA), Central Institute of Freshwater Aquaculture (CIFA), CMFRI, and others have taken necessary steps for conducting various training programs, field-level extension programs, lab-to-land programs, operational research projects, and entrepreneurship development programs to increase the technical knowledge of farmers, disease prevention, and improvement of management. These principal fishery research institutes have also prepared a good number of printed extensions and training materials in local regional languages for the local people.

In India, the Ministry of Agriculture issued guidelines for sustainable development and management of brackish water aquaculture whose overall purposes were to assist in the formulation of appropriate shrimp farming management practices, adopting measures for mitigating the environmental impacts of shrimp pond waters, and judicious utilization of land/water resources. The Aquaculture Authority of India was set up when the Supreme Court of India passed a ruling that prohibited the setup or construction of shrimp culture ponds within the coastal regulation zone (CRZ). This body formulated guidelines for adopting technologies and optimizing yield levels on a sustainable basis in traditional and modern aquaculture systems. It has been made mandatory that all shrimp farms of 5 hectares of water area within the CRZ and 10 hectares outside the CRZ should have an effluent treatment system. The establishment of such a system is necessary to bring shrimp farm wastewater within the prescribed standard and mitigate any adverse impacts on the ecology of the open waters. The Aquaculture Authority has also constituted state-level committees (SLCs) and district-level committees (DLCs). Applications submitted by farmers are received by the DLCs. After verification of the information and field-level inspections, wherever necessary, the applications are forwarded to the SLCs for consideration. After the recommendation of the SLC, the applications are forwarded to the Aquaculture Authority for approval. Any approval will be given in Form IV for three years and will be subject to certain conditions.

The National Centre for Sustainable Aquaculture (NaCSA) was established by MPEDA in 2007 as an outreach organization for uplifting the livelihood of small-scale shrimp farmers. NaCSA started grouping these farmers into societies and educated them on better management practices (BMPs) for safe and sustainable shrimp farming. NaCSA also trained these farmer societies to follow a cluster approach in shrimp farming. NaCSA mainly functions as the prime mover of the extension activities among aquaculture farmers. It organizes small-scale farmers farming within the same creek into societies and promotes better management practices to improve their production and profit. It conducts meetings on crop planning and technical updates apart from this it technical support to society and non-society farmers during culture to allow them to market their products to quality-demanding export markets.

The government of India launched the Pradhan Mantri Matsya Sampad Yojana (PMMSY) on September 10, 2020, to transform the fisheries sector and add strength to efforts of building an *Aatmanirbhar Bharat*. The PMMSY is a flagship scheme for focused and sustainable development of the fisheries sector in the country, with an estimated investment of Rs20,050 crore. The scheme is aimed to support farmers with various assistance, and hence it is envisaged that the scheme will help double the income of farmers.

The role of the local government is paramount in the development of sustainable shrimp farming. The local government body has to develop a complete sense of ownership of the development process and move ahead with shrimp farmers in a participatory model. Motivating farmers on the use of good management practices and awareness building has been a herculean task for which the present framework in the local government has proven inadequate. This has been compounded by the absence of qualified NGOs with work experience in the fisheries sector, especially in shrimp farming. To overcome this lacuna, it is emphasized that shrimp farmers should organize into associations or self-help groups. The formation of "aqua clubs" in some shrimp farming areas in various states of India is a significant step, and state governments should promote and further this movement to include all shrimp farmers.

8.4 CONCLUSION

Shrimp is an international product, and promotion should take place at an international level. Producers worldwide should therefore collaborate through their representative bodies (e.g., EMPEDA) to promote shrimp and initiate quality labeling initiatives and guarantees to consumers of environmentally sound production practices.

Current production could be increased many times over by implementing proper development strategies, engineering expertise, and technical know-how for the development of land and water areas, and also the enormous labor force required for the purpose is available in India. Research institutes have also developed the technology for the mass production of shrimp seed and feed. Extension and financing services for shrimp farming are also available with different organizations. Side by side with the fast and intensive expansion of shrimp farming activities, various problems also are cropping up. Environmental problems like pollution, localized problems of a socio-economic nature, and the recent disease problems in hatcheries and farms alike call for serious attention.

The Indian market continues to grow well and is less sensitive to an environmental awareness. In terms of production effects on price, diseases are likely to continue to cause problems in many areas and act as a brake on excessive production and consequential reduced price. This is an interesting example of a sustainability conundrum: lack of sustainability among some producers may secure sustainability for others. Production from wild fisheries is likely to stay steady or decline.

8.4.1 Acknowledgments

The authors gratefully acknowledge the help from Vivekananda Maity, Fishery Extension Officer, West Bengal, India, in preparing the manuscript.

REFERENCES

Alagarswami K (1981) Prospects for coastal aquaculture in India. In: Proc. Seminar on Role of Small Scale Fisheries and Coastal Aquaculture in Integrated Rural Development, CMFRI Bulletin, Madras, December 6–9, 1978, pp. 30-A.

Arnold WS, White MW, Norris HA, Berrigan ME (2000) Hard clam (*Mercenaria* spp.) aquaculture in Florida, USA: geographic information system application to lease selection. Aquacult. Eng. 23, 203–231.

Brundtl and Commission Report (1987) Retrieved October 2, 2008 from http://worldinbalance.net/agreements/1987-brundtland.php.

George MJ, Mohamed KH, Pillai NN (1968) Observations on the paddy field prawn filtration of Kerala, India. FAO Fish. Rep. 57(2), 427–442.

Giap DH, Yi Y, Yakupitiyage A (2005) GIS for land evaluation for shrimp farming in Haipong of Vietnam. Ocean Coast. Manage 48, 51–63.

Gordon HR, Morel A (1983) Remote Assessment of Ocean Color for Interpretation of Satellite Visible Imagery: A Review. Springer-Verlag, New York, 144 pp.

Gupta MC (1995) Brackish water aquaculture site selection using techniques of geographical information system (GIS). Space Application Centre ISRO, Ahmedabad.

Hein L (2000) Impact of shrimp farming on mangroves along east coast of India. Unasylva 51, 48–54.

Hossain MZ, Muttitanon W, Tripathi NK, Philips M (2003) Monitoring shrimp farming development from the space. Gisdevelopment 7, 1–6.

Jayanthi M, Rekha PN (2004) Role of remote sensing and geographic information system for natural resource assessment in brackish water aquaculture. In: Proceedings of the Natural Resource Management and Agro-Environmental Engineering, International Conference on Emerging Technologies in Agriculture and Food Engineering, IIT Kharagpur, India, December 14–17, pp. 41–48.

Karthik M, Suri J, Sahara NN, Birdar RS (2005) Brackish water site selection in Palghar Taluk of Thane district of Maharashtra, India, using remote sensing and GIS. Aquacult. Eng. 32, 285–302.

Kumaran M, Ponnusumy K, Kalaimani N (2003) Diffusion and adoption of shrimp farming technologies. Aquacult. Asia 8, 20–23.

Manoj V, Vasudevan N (2009) Functional options for sustainable shrimp aquaculture in India. Rev. Fish. Sci. 17(3), 336–347.

Menon MK (1954) On the paddy field prawn fishery of Travancore-Cochin and an experiment in prawn culture. Procceeding of Indo-Pacific Fish Counc. Sth Session, Section II, pp. 1–5.

MIDAS (Mekong International Development Associates) Agronomic (1995) Pre-investment study fora coastal resources management program in Thailand. Final report for the World Bank and the Office of Agriculture Economics. Ministry of Agriculture and Cooperatives, Bangkok.

Muthu MS (1978) A general review of penaeid prawn culture: Summer Institute in Breeding and Rearing of Marine Prawns, CMFRI Special Publication No. 3, India, pp. 25–33.

NACA (Network of Agriculture Centres in Asia-Pacific) (1969) A survey project for the sources of waste water pollution from coastal fishery. A Research Report Submitted to the Department of Pollution Control, Ministry of Science, Technology, and Environment, Bangkok.

Pillai CT (1984) Handbook on diagnosis and control of bacterial diseases in finfish and shellfish culture. CMFRI Sp. Pub. No. 17, India, pp. 1–51

Rajitha K, Mukherjee CK, Chandran RV (2007) Applications of remote sensing and GIS for sustainable management of shrimp culture in India. Aquac. Eng. 36, 1–17.

Ramamurthy S (1978) Prawn farm. In: Summer Institute in Breeding and Rearing of Marine Prawns, May 11-June 9, 1977, CMFRI Spl Pub. No. 3, Cochin, pp. 92–103.

Ramesh R, Rajkumar R (1996) Coastal aqua culture site selection and planning in Tamil Nadu using remote sensing and GIS. Asian-Pacific Remote Sens. GIS J. 9(1), 39–49.

Rao PV (1981) Recent technological advances in coastal aquaculture in India. In: Proc. Seminar on the Role of Small-Scale Fisheries and Coastal Aquaculture in Integrated Rural Development, CMFRI Bulletin, Madras, December 6–9, 1978, p. 30-A.

Rao PV (1983) Studies on penaeld prawn diseases. In: Summer Institute in Hatchery Production of Prawn Seed and Culture of Marine Prawns, April 18–May 17, 1983, Central Marine Fisheries Research Institute, Cochin, India.

Salam MA, Ross LG (2000) Optimizing site selection for development of shrimp (*Penaeus monodon*) and mud crab (*Scylla serrata*) culture in Southwestern Bangladesh. In: 14th Annual Conference on Geographic Information System, Proceeding of the GIS'2000, Toronto, Canada, March 13–16.

Satyanarayan B, Raman AV, Frank D, Kalavati C, Chandramohan P (2002) Mangrove floristic and zonation patterns of Coringa, Kakinada Bay, East Coast of India. Wetlands Ecol. Manage. 10, 25–39.

Seethalakshmi PS, Rajeev R, Kiran SG, Selvin J (2021) Shrimp disease management for sustainable aquaculture: innovation from nanotechnology and biotechnology. Aquac. Int. 29, 1591–1620.

Silas EG (1978) Research and development programme in the culture and progapation of marine penaeid prawns. Summer Institute in Breeding and Rearing of Marine Prawns CMFRI Special Publication No. 3, pp. 17–25.

Tong PHS, Auda Y, Populus J, Aizpuru M, Habshi AAL, Balsco F (2004) Assessment from space of mangroves evolution in the Mekong delta, in relation to extensive shrimp farming. Int. J. Remote Sens. 25, 4795–4812.

Unnithan KA (1985) A Guide to Prawn Farming in Kerala. CMFRI Spl. Pub. No. 21.

Unnithan KA (2000) Shrimp farming—a status review. In: *Marine Fisheries Research and Management*, V.N. Pillai and N.G. Menon (eds), CMFRI Bulletin, India, pp. 727–746.

Vadlapudi S (1996) Identification and quantification of changes in mangrove forest using remote sensing in the Kakinada Bay, Andhra Pradesh, India. GIS Dev., 1–9.

Wurts AW (2000) Sustainable aquaculture in the twenty-first century. Rev. Fish. Sci. 8, 141–150.

9 Biochar as a Potential Candidate for Removal of Pollutants from Aquaculture

Thanh Binh Nguyen, Xuan-Thanh Bui, Quoc-Minh Truong, Thi-Kim-Tuyen Nguyen, Thi-Bao-Chau Ho, Phung-Ngoc-Thao Ho, Van-Re Le, Van-Anh Thai, Hien-Thi-Thanh Ho, Van Dien Dang and Cheng-Di Dong

CONTENTS

9.1 Introduction	121
9.2 Biochar Production	122
9.2.1 Synthesis Method	122
9.2.1.1 Pyrolysis	122
9.2.1.1.1 Conventional Pyrolysis	122
9.2.1.1.2 Slow Pyrolysis	122
9.2.1.1.3 Fast Pyrolysis	122
9.2.1.1.4 Flash Carbonization	123
9.2.1.2 Torrefaction	123
9.2.1.3 Hydrothermal Carbonization	124
9.2.2 Biochar Characteristics	124
9.2.2.1 Physical Properties	124
9.2.2.2 Chemical Properties	125
9.3 Water Remediation	126
9.3.1 Adsorbents	126
9.3.2 Catalyst	126
9.3.3 Capacitive Deionization	128
9.4 Conclusions and Future Research Needs	129
References	129

9.1 INTRODUCTION

A wide range of biological and physicochemical approaches have been developed to deal with polluted water and wastewater (W&W) and to meet the demanding environmental restrictions of industry. The activated sludge process and anaerobic digestion are low-cost and efficient methods for dealing with low- to moderate-strength effluents. Sludge management costs and inadequate removal of some non-biodegradable contaminants, which are of growing concern, are two constraints. This has led to the development of a wide range of treatment technologies, including ultrasonic and microwave radiation, Fenton-like oxidation, and catalytic decomposition of organic pollutants (especially with advanced engineered nanomaterials) via advanced oxidation processes (AOPs). The elimination of organic and inorganic substances can also be achieved by adsorption techniques. Activated carbon is the most common substance used.

DOI: 10.1201/9781003326946-9

One way to protect public health and prevent water-borne disease epidemics in disadvantaged communities is to use low-cost water treatment. Using locally available adsorbent materials, high efficiency (>90%) and selectivity, cost effectiveness, and keeping flavor and color are all advantages of adsorption-based contaminant removal. Converting food production wastes (such crop residue) into biochar by pyrolysis and then using it for water treatment has promising but underutilized potential. Low-cost biochar, a carbon-rich solid generated by pyrolysis of biomass, has received considerable international scientific attention. There has been a lot of research on the use of biochar as a soil supplement to improve soil quality, reduce greenhouse gas emissions, and sequester carbon. Research on biochar's ability to remove contaminants from aqueous solutions is based primarily on batch tests, and there has been little research on the design and optimization of biochar-based systems for drinking water and wastewater treatment. Depending on the feedstock and the pyrolysis conditions, the advantages of employing biochar vary.

The current analysis offers an overview of worldwide efforts to produce biochar for wastewater treatment applications and analyze the technology against the sustainability criteria in the context of this review. High-quality reviews published in recent years and highly referenced papers have been identified as current trends. For future research, these findings give a critical overview and identify existing knowledge gaps that can be used to guide future research in this topic.

9.2 BIOCHAR PRODUCTION

9.2.1 Synthesis Method

9.2.1.1 Pyrolysis

Pyrolysis is one of the most promising methods for transforming biomass into primary biofuels and biochars due to its speed and ease of usage. Biochar yield, characteristics (porous or amorphous), and quality are significantly influenced by the manufacturing process parameters (temperature, residence time, heating rate, pressure, chemical composition, shape, and size). Additionally, the content, structure, and intrinsic binding of the original biomass influence the physicochemical properties of biochar. Biochar production during pyrolysis is influenced by variables such as biomass moisture content and particle size. The pyrolysis process is classified into two forms, conventional pyrolysis and microwave-assisted pyrolysis, based on differences in operating conditions.

9.2.1.1.1 Conventional Pyrolysis

Conventional pyrolysis is the process of heating biomass under anaerobic conditions at a particular temperature range between 300 and 700°C, which can be used to produce biochar. Pyrolysis is classified as slow, fast, or flash based on vapor residence time, temperature, and heating rate. The enhanced residence time, lower pyrolytic temperature, and reduced heating rate all lead to increased biochar production.

9.2.1.1.2 Slow Pyrolysis

Compared to gasification and rapid pyrolysis, slow pyrolysis generates greater biochar outputs from diverse biomass sources. Slow pyrolysis is characterized by a sluggish rise in temperature and a slow heating rate. Upon completion of the process, substantial quantities of biochar and smaller quantities of liquid and gaseous products are obtained. At various stages of biomass digestion, temperatures between 400 and 500°C are employed. First, bonds are broken and moisture is extracted, followed by the degradation of carbohydrates, lipids, and proteins, and finally, carbonaceous residues are produced.

9.2.1.1.3 Fast Pyrolysis

Fast pyrolysis is typically conducted between 450 and 600°C with a higher heating rate (200°C min^{-1}) and shorter residence time (a few seconds) than slow pyrolysis. Due to the brief duration of

Biochar and Removal of Pollutants from Aquaculture

the process, the effects of heat and mass transfer, dynamics, and other variables have a substantial impact on product yield and process efficiency. The low biochar output (10–20%) is the result of operating circumstances including fast pyrolysis. Rapid pyrolysis may have created biochar with a low calorific value and a high oxygen content due to the short residence time.

9.2.1.1.4 Flash Carbonization

Flash carbonization provides a higher biochar yield (28–32%) and a shorter reaction time (less than 30 min), making it a more efficient biochar production method than with classical carbonization. The feedstock is packed into a packed bed reactor in the flash carbonization process. The vessel is then pressurized with air to a pressure of 1–2 bar, and the bottom of the pressurized vessel is heated with a flame. The flame was moved upwards by the downstream airflow, which took less than 30 min to heat up the entire packed bed (Kumar et al., 2020). In general, a specific level of pressure is required for flash carbonization.

9.2.1.2 Torrefaction

Torrefaction is a thermochemical conversion that occurs in the absence of oxygen at atmospheric pressure and temperatures between 200 and 300°C with a modest heating rate (less than 50°C/min) and a relatively long residence time (20–120 minutes). The process creates a solid product with a high carbon content, known as torrefied biomass or char, by partially decomposing biomass. Torrefaction is a new thermal biomass pre-treatment process that eliminates volatiles from biomass by a variety of decomposition events, hence enhancing biomass quality and modifying combustion behavior. By partially decomposing biomass, this method yields a solid product with a high carbon content known as torrefied biomass or char. Torrefaction is a novel thermal biomass pre-treatment method that eliminates volatiles from biomass through a variety of decomposition events, hence improving biomass quality and altering combustion behavior. The hydrophobicity of biochar produced from surface functional groups and its decreased ash content are crucial characteristics that influence its efficacy in soil and water contamination absorption. Torrefaction is often performed at a low temperature, with a brief residence period and a moderate heating rate, in order to provide a higher yield of solid products. When the torrefaction temperature was increased, the yield of solids decreased. The effect of residence time on the mass yield of torrefied biomass at 300°C revealed that the mass yield decreased as the residence time rose. Approximately 30% of the mass of certain highly reactive volatile compounds is transformed into torrefied vapor during this process (Ma et al., 2019). Torrefied biochar can have a similar energy density to coal (22 MJ/kg), which is used for heating and electricity production (Phanphanich & Mani, 2011).

Typically, in order to provide energy for the torrefaction process, the torrefied volatiles are immediately burned in a gas combustor. Torrefaction demands a high torrefaction temperature and a long residence time in order to produce biochar with a high energy density, which affects the quality and energy output of torrefied biochar. To obtain biochar with a relatively greater heating value, mass energy density, and energy yield, the ideal biomass torrefaction condition may be a solid yield between 60 and 80%. The quality of torrefied biochar is affected by the physical-chemical characteristics of the biomass, such as its moisture content, higher heating value, and ash content (Medic et al., 2012). The moisture content should be the most essential since it controls the majority of the energy input during the torrefaction process. Dehydration and decarboxylation are the primary decomposition processes that contribute significantly to torrefaction mass loss.

At such a low torrefaction temperature, the feedstock can be simply dried without undergoing significant chemical processes. In spite of this, the benefits of torrefied biochar continue to get considerable interest. Torrefied biomass may be processed into fine powders for use in pulverized coal-fired power plants more easily than raw biomass resources (Barskov et al., 2019).

9.2.1.3 Hydrothermal Carbonization

Hydrothermal carbonization (HTC) is one of the hydrothermal processes, which is a thermochemical conversion technique that employs heat to convert wet biomass feedstocks into gaseous, aqueous compounds, and hydrochar. HTC is a low-cost and environmentally friendly technology with benefits such as biomass use for HTC without the need for drying, a benign atmosphere, a low toxicological impact of materials, a reduction in ash content, and inorganic elemental composition. German chemist Friedrich Bergius discovered HTC in 1913. He observed that biomass feedstock using water as a carbonization medium, which performed in an oven at temperatures ranging from 180 to 250°C and under autogenous pressure, decreased the oxygen and hydrogen content of the feedstock based on five major reaction mechanisms: hydrolysis, dehydration, decarboxylation, polymerization, and aromatization. Hydrolysis, which breaks the ester and ether bonds of hemicellulose, cellulose, and lignin at varying temperatures, is the initial step. The dehydration of a biomass matrix will release water into the water phase. On the basis of carboxyl and carbonyl groups, the decarboxylation mechanism will produce CO_2 and CO. Consequently, dehydration and decarboxylation are the primary causes for a decrease in the H/C and O/C ratios, as they increase the weight percentage of carbon and the hydrophobicity of the final carbon substance. Small molecules join to produce a bigger molecule during the polymerization reaction. The last aromatization reaction is caused by dehydration and decarboxylation, which is the heart of hydrothermal carbonization since it leads to the creation of aromatic rings on the hydrochar surface. In the hydrothermal carbonization process, these processes occur concurrently and are interrelated.

HTC is governed by temperature, residence time, heating rate, reactant concentration, and the water-to-raw-biomass ratio, with reaction temperature being the most influential variable on hydrochar characteristics. Water is a crucial substance because it serves as an active transfer channel for ions and rearranges the structure of biomass. Moreover, the content and structure of the original biomass can influence the physical features of hydrochar, such as its structure, shape, surface morphology, and size.

Hydrochar, the primary byproduct of carbonization, has a high carbon content, high hydrophobicity, an abundance of oxygen-containing functional groups on its surface (aromatic structure), and a microporous or mesoporous structure, which makes them suitable materials as precursors of activated carbon and may enhance the ability of hydrochar to remove pollutants from wastewater. Hydrochar is a homogenized carbon-rich, carbon-dense material with a better water-holding ability than unprocessed biomass. Depending on the feedstock and HTC reaction conditions, hydrochar contains 50–90% of the dry mass of the raw biomass and 70–90% of its original energy content. In recent years, bio-products formed by the HTC process has the potential for a variety of applications as a soil amendment, bioenergy feedstock (including liquid fuel or bio-oil), adsorbent for contaminant (Libra et al., 2011), and carbon catalyst for the production of chemicals (Wataniyakul et al., 2018). With the recent development of HTC technology, it has been proved that biomass and waste materials have the potential to produce a number of products that will be employed in a variety of applications in the future and that these applications will then extend to practical uses.

9.2.2 BIOCHAR CHARACTERISTICS

9.2.2.1 Physical Properties

Bulk density is the volumetric specific weight of bulk material in a heap, encompassing voids in the solid structure and spaces between various pieces of the mass. Due to the escape of volatile chemicals and the creation of graphite crystals, the biochar bulk density is typically larger than that of the biomass feedstock. The increasing pyrolysis temperature increases the bulk density of biochar, but it is not dependent on the heating rate, resulting in a greater reduction of the solid substrate and intensity of carbonization. The bulk density of greenwood biomass reduced from 700 kg/m³ to approximately 400 kg/m³ after drying at 40°C and then further decreased to 300 kg/m³ after pyrolysis at 500°C at a heating rate of 10°C/min (Abdullah & Wu, 2009).

To evaluate the possible application of biochar, the pore size distribution is one of the most important physical features. According to the diameter of the inner wall of big spores (>50 nm), mesopores (2–50 nm), and microspores (2 nm), the pore diameters of biochar can be characterized (Rouquerol et al., 2013). Among the pore sizes of biochar, micropores are the major contributors to the biochar surface area and efficient adsorption of molecules. The micropore volume of biochar can be enhanced by increasing the pyrolysis time of the biomass at higher temperatures. The specific surface area of biochar can be massively increased through increasing pyrolysis temperature due to disruption of the pore walls, causing expansion of the micropores. Pore enlargement can reduce micropore volume but increase total pore volume and biochar surface area. The porosity variations due to the release of water and volatile gases during the pyrolysis of biomass contribute to the formation and development of the biochar pore structure. The biochar specific surface area is closely related to its porosity, with micropore and mesopore volume being two main parts that contribute to biochar pore volume. The microparticle volume accounted for about 12.1–58.0% of the total biochar pore volume, while the mesopore volume fraction varied from 18.9 to 31.7%. The pore volume and surface area can be examined using the Brunauer, Emmett, and Teller (BET) technique, where N_2 adsorption–desorption isotherms are the most applied sorbate gas. However, for micropore analysis, N_2 gas is not always suitable to provide accurate results because of some limitations, such as N_2 concentrating in the micropores, preventing absorption and replacement at low temperatures (77K). Instead, CO_2 showed more accurate results, and it can be used at high temperatures (273K).

Mechanical strength plays a key role in determining the ability of biochar to withstand tear and abrasion stress in environmental applications. The mechanical stability is significantly related to density and therefore inversely related to porosity. Compared with rapid pyrolysis, low pyrolysis can improve the compressive strength, which may explain the faster evaporation of water and volatile matter when the heating rate is fast. Therefore, pyrolysis of original biomass feedstock that has lower humidity contents produces biochar with higher mechanical strength.

9.2.2.2 Chemical Properties

The surface functional groups play a key role in the properties of biochar to estimate the potential feasibility of the application of biochar in various fields such as adsorbents, electrodes, and catalysts. The composition of biochar is heterogeneous. For example, the O, N, P, and S heteroatoms are combined with the aromatic structure of the biochar, resulting in chemical inhomogeneity due to the electronegativity difference between the biochar. The heterogeneous composition gives diverse and wide-ranging surface chemistry and exhibits hydrophobic and hydrophilic functional groups and basic and acidic properties based on production conditions and biomass. A variety of distinct S and N functional groups have been identified in biochar, particularly those functional groups originating from municipal sewage and manure (Leng et al., 2020). The functional groups of biochar can be categorized as electron acceptors and electron donors.

In addition to a thorough understanding of the characteristics of biochar functional groups, their dynamic thermal behavior under diverse reaction conditions should be carefully considered. The thermal degradation of lignocellulose is transformable by dehydration, pyrolysis, polyaromatic graphene sheets, and carbonization at 250, 350, and 600°C, respectively. The loss of O, H, and C in the form of H_2O, hydrocarbons, CO_2, CO, H_2, and tar vapor mainly contributes to the loss of biomass during dehydration and pyrolysis. Typically, C substance enhances by 40–50 wt% in biomass feedstock to 70–80 wt% in the manufactured biochar after pyrolysis at 250–600°C. Moreover, the content C of biochar can reach higher than 90 wt% through carbonization.

The main components of biochar are C, H, and O, and its content is highly variable depending on the feedstock. Typically, the carbon substance of biochar varies from 45 to 60% by weight, while hydrogen and oxygen substances are in the range of 2–5% and 10–20% by weight, respectively. In addition to these components, other inorganic compounds present in biochar have a substantial impact on its characteristics. Inorganic compounds in biomass may be evaporated, entrained with charcoal, or maintained as a distinct mineral phase. The quantity and composition of inorganic materials are largely dependent on the biomass input. Typically, wood biomass, including wood chips and sawmill byproducts, has a low ash percentage (1 wt%). Straw, rice husk, and grass have a high ash percentage (up to 24 wt%) because their silica level is high. The temperature of pyrolysis also affects the inorganic element content of biochar. Inorganic contents such as P, Mg, K, Ca, Al, and Fe of biochar generated from wheat straw obviously increased due to the increased pyrolysis temperature. In general, the minerals intrinsic to biomass can affect the properties of biochar by interacting with its organic constituents during pyrolysis.

9.3 WATER REMEDIATION

9.3.1 Adsorbents

Adsorbents were evaluated based on criteria such as capacity, selectivity, reproducibility, kinetics, compatibility, and cost. However, it is very difficult to meet all these requirements. It is possible to narrow the criteria for adsorbent selection to one or two. To evaluate whether the selected adsorbent material matches the set criteria, information can be found in many different ways. For common compounds, product information may be provided by the manufacturing company. Previous studies can also be consulted. Compound information may also be self-collected or collected by another independent company.

The most important property of an adsorbent is the adsorption capacity. The amount of adsorbent that occupies the surface of the adsorbent per unit mass or volume is called the adsorption capacity. External factors such as temperature, adsorbent concentration, and pH also contribute to the adsorption capacity. To evaluate the adsorption capacity of the adsorbent in the most general way, the data were collected under constant temperature conditions. The data were calculated according to the isothermal equation with different adsorbent concentrations. The reproducibility of the material is also a feature of interest for adsorbents. However, methods to make compounds reusable have a low cost requirement. Regeneration methods include thermal swing, pressure swing, or using chemicals, or a combination of these may be used to increase regeneration efficiency. In fact, the ability to regenerate will gradually decrease over reuse. The cycle time of the adsorption process is controlled by mass transfer kinetics. When equilibrium is reached, the adsorption curve will tend to draw a horizontal line. It is possible to prolong the time to reach equilibrium by increasing the adsorbent concentration or decreasing the amount of adsorbent (decreasing the flux per adsorbent unit). The compatibility of the adsorbent is assessed by chemical or physical effects that reduce its life. The medium reacts with external influences including temperature, pressure, and vibration which can potentially lead to the decomposition of the adsorbent. The cost of adsorbents also determines their applicability in practice. However, not all low-cost adsorbents can be highly effective. Therefore, there have been many studies to improve the efficiency of adsorbents.

Currently, to treat polluted water sources, researchers have shown that using adsorbents to remove toxic substances is the most effective. Previously studied adsorbents include activated carbon, chitosan, zeolites, clay minerals, plant-based materials, and industrial waste. Biochar is evaluated as a potential adsorbent because of characteristics such as low cost, a simple preparation method, high adsorption efficiency, and reproducibility. Wastewater treatment systems containing biochar, the efficiency of organic pollutants is improved compared with no biochar one. Table 9.1 summarizes previous studies using biochar as an adsorbent in water treatment.

9.3.2 Catalyst

Catalysts have an effect on the chemical reaction process by definition. Only the rate of the reaction is accelerated, not its nature. Even in the presence of a catalyst, the reaction produces more products. Deactivation decreases and selectivity improves under particular conditions, such as low temperature and pressure. Carbon-based materials are employed as catalysts in industrial applications to facilitate the most effective reactions. In recent years, biochar has emerged as a viable option to

TABLE 9.1
Adsorption Capacities of Different Biochars for removing Organic Pollutants from Water

Type of Biochar	Adsorbed	Adsorption Capacity (mg g^{-1})	References
Durian shells	Cd(II)	37.64	(Yang et al., 2021)
Branches of *Robinia pseudoacacia*		11.37	
Pomelo peel	Cr(VI)	24.37	(Dong et al., 2021)
	Phenol	39.32	
Sewage sludge	Triclosan	67.9–100	(Czech et al., 2021)
	Diclofenac	88.7–1410	
	Naproxen	10.1–135	
Chinese herbs	Tetracycline	188.7	(Wang et al., 2019)
	Chlortetracycline	200.0	
	Oxytetracycline	129.9	
Mixed food scraps and plant trimmings	Tetracycline	15.52	(Hoslett et al., 2021)

replace carbon-based materials due to its high porosity and carbon content. Figure 9.1 summarizes biochar's role as a catalyst in biodiesel generation, tar removal, NO$_x$ removal, and biooil production. The origin of biomass, conditions of biochar production, and other factors all have an impact on the properties of biochar. Coal processing must therefore be carried out in optimal conditions in order to maximize its qualities.

When biochar is used as an adsorbent, contaminants are simply transported from one medium to another. As a catalyst, it can be transformed into byproducts that are less hazardous and more biodegradable. The redox-active groups, such as quinone, hydroquinone, and conjugated n-electron systems, present on the surface of biochar enable it to act as an electron donor for dissolved oxygen. In Fan et al. (2021), BPA removal efficiency was raised to 94.5% by using biochar generated from sludge as a catalyst to activate peroxymonosulfate (PMS) to remove bisphenol A in water. Another study on PMS activation when catalyzed with coconut shell-derived biochar by Hung et al. (2022) showed that sulfamethoxazole was degraded to 99% in chloride-rich systems.

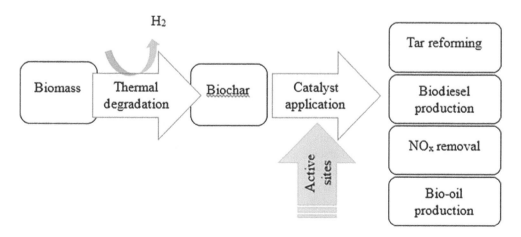

FIGURE 9.1 Application of biochar as a catalyst.

9.3.3 CAPACITIVE DEIONIZATION

Capacitive deionization (CDI) is a promising electrochemical method that uses porous electrode materials to electrosorb undesirable ions from aqueous solutions. Further integrating CDI with bioelectrochemical systems (such as microbial fuel cells) could provide emerging technologies for wastewater treatment. Among the several advantages of a CDI system are its low energy consumption, high water recovery, direct energy recovery, lack of chemical additives, limited fouling potential, and environmentally friendly operation. A porous electrode material can have a substantial effect on electrosorption in CDI. The use of porous carbon materials, such as activated carbon, for CDI has been extensively researched, carbon aerogels, carbon fibers, carbon nanotubes, graphene, templated mesoporous carbon, and composites of these materials. Although promising electrode materials for CDI have been developed, a significant challenge is the high cost and complicated preparation methods needed to implement them at a large scale. Electrode material accounts for approximately 60–70% of the overall cost of carbon-based EDL capacitors. Most reported mesoporous carbon materials (e.g., OMC, nanotubes, modified carbon aerogels) need complex polymerization procedures and costly polymeric substrates, resulting in high costs and secondary environmental damage (Burke, 2007; Xu et al., 2008). As a result, attempts to solve these issues, together with good electrode design, are critical for advancing CDI technology.

The production of porous carbon from low-cost biomass leftovers (e.g., wood, tree barks, agricultural residues, and fruit bunches) has received considerable attention in recent years. Along with their economic advantage, their renewable nature and low secondary environmental effect make them more viable options than carbon produced from polymeric substrates. One alternative in this respect is biochar (i.e., a by-product of biomass pyrolysis), which is also being investigated as a carbon sequestration option. Activated biochar is considered a potentially renewable carbon-based electrode for energy storage (supercapacitors) and wastewater treatment (electrosorption), for example, sugarcane bagasse fly ash (SCBFA), a major biowaste with a high fixed carbon content, as a precursor for the production of activated carbon (SCBFA-AC). SCBFA-AC was valorized as the primary component of carbon electrodes used in capacitive deionization, a new desalination technology. CDI and membrane CDI experiments demonstrated the critical role of ion-exchange membranes in increasing charge efficiency (from 5–30 to 80–95%) and hence salt adsorption capacity (SAC) from 5 to 22 mg g^{-1}. After 70 CDI cycles, this SAC value, which is one of the highest ever recorded with biowaste electrodes, showed just a tiny decrease (19 mg g^{-1}). SCBFA electrodes were successfully evaluated for water softening applications, with a maximum of 15 mg g^{-1} when $CaCl_2$ solutions were used instead of NaCl (Lado et al., 2019). Adorna et al. (2020) developed a Faradaic/electric double-layer hybrid capacitor for capacitive deionization applications using activated biochar (AB) derived from coconut shells and a MnO_2 nanocomposite. The α-MnO_2 was successfully deposited on the AB, resulting in an excellent electrode material for CDI applications using neutral Na_2SO_4 electrolytes. Notably, AB–MnO_2 nanocomposites prepared via the indirect co-precipitation method (AB–MnO_2-indirect) have a large specific surface area of 304 m^{2}g^{-1}, a high mesopore volume ratio, a high capacitance retention, a high hydrophilicity, and an appropriate pore texture for reducing the ion diffusion distance. Additionally, AB–MnO_2-indirect exhibits exceptional CDI performance at 1.2 V, exhibiting a specific electrosorption capacity of 33.9–68.4 mg g^{-1} (Adorna et al., 2020). For arsenic-contaminated groundwater remediation, Cuong et al. (2022) used combined filtration and electrosorption. An active rice hulk biochar filter preoxidized As(III) to As(V) and removed As (III, V), whereas CDI post-treated arsenic and other ions in groundwater. The combined system has considerable potential for neutralizing As (III). The efficient redox transition of As(III) to As(V) resulted in 94% oxidation after 60 hours. The performance of the active BC filter improved with starting arsenic content and decreased flow rate. The arsenic electrosorption capability rose from 0.4 to 1.2 V in CDI. The AC electrodes may be double-layer charged to eliminate most As(V). A combined system with active rice hulk biochar filter and CDI was also used to remediate actual groundwater. These amounts in the effluent met WHO guidelines for drinking water quality, proving that the system can remediate other chemicals as well as arsenic (Cuong et al., 2022). Zhaoyang Du et al. (2020) used red oak bio charcoal, a new electrode material that is both affordable and highly conductive, to perform capacitive deionization on leather production effluent to concurrently remove Cl and Cr^{3+}. The ideal working settings (voltage of 1.8 V, flow rate of 10 ml/min, and electrode spacing of 2 mm) resulted in a maximum adsorption capacity of 13.79 mg/g for the KOBC15 electrode. After 10 cycles of CDI system operation, the removal rates of Cl and Cr^{3+} in simulated leather production effluent were 86.7% and 100%, respectively (Du et al., 2020). However, when compared to previous studies, hierarchical porous carbon prepared from biochar needed to improve in terms of charge efficiency and other performance. Despite this, biochar still demonstrated the potential to achieve high-performance electrosorption in the wastewater treatment industry. Furthermore, CDI is often used for sustainable desalination of brackish water. It is fair to expect that hierarchical porous carbon created by biochar would improve competitiveness in both cost and energy.

9.4 CONCLUSIONS AND FUTURE RESEARCH NEEDS

The available findings of biochar research strongly justify the potential of environmental benefits and limitations of biochar for expanding its use in remediation of water. Biochar has enormous potential for the removal of pollutants such as organic and inorganic contaminants from wastewater. However, one type of biochar may not be appropriate for all contaminant removal. Pyrolysis condition and feedstock type are the main factors influencing the applicability of biochar. In summary, greater emphasis should be placed on practical applications, and additional research should be conducted to address existing issues. (1) Current research on the stability of BC or BC-based composites and their biological toxicity to aquatic and soil microorganisms is lacking; (2) the versatile area of biochar application needs to be promoted for real wastewater application; and (3) the mechanism by which BC removes contaminants is still inconclusive, and the relationships between biomass, preparation methods, and properties are still unclear.

REFERENCES

Abdullah, H., Wu, H. 2009. Biochar as a fuel: 1. Properties and grindability of biochars produced from the pyrolysis of mallee wood under slow-heating conditions. *Energy Fuels*, **23**(8), 4174–4181.

Adorna, J., Jr., Borines, M., Doong, R.-A. 2020. Coconut shell derived activated biochar–manganese dioxide nanocomposites for high performance capacitive deionization. *Desalination*, **492**, 114602.

Barskov, S., Zappi, M., Buchireddy, P., Dufreche, S., Guillory, J., Gang, D., Hernandez, R., Bajpai, R., Baudier, J., Cooper, R., Sharp, R. 2019. Torrefaction of biomass: A review of production methods for biocoal from cultured and waste lignocellulosic feedstocks. *Renew. Energy*, **142**, 624–642.

Burke, A. 2007. R&D considerations for the performance and application of electrochemical capacitors. *Electrochim. Acta*, **53**(3), 1083–1091.

Cuong, D.V., Wu, P.-C., Liou, S.Y.H., Hou, C.-H. 2022. An integrated active biochar filter and capacitive deionization system for high-performance removal of arsenic from groundwater. *J. Hazard. Mater.*, **423**, 127084.

Czech, B., Konczak, M., Rakowska, M., Oleszczuk, P. 2021. Engineered biochars from organic wastes for the adsorption of diclofenac, naproxen and triclosan from water systems. *J. Clean. Prod.*, **288**, 11.

Dong, F.X., Yan, L., Zhou, X.H., Huang, S.T., Liang, J.Y., Zhang, W.X., Guo, Z.W., Guo, P.R., Qian, W., Kong, L.J., Chu, W., Diao, Z.H. 2021. Simultaneous adsorption of Cr(VI) and phenol by biochar-based iron oxide composites in water: Performance, kinetics and mechanism. *J. Hazard. Mater.*, **416**, 11.

Du, Z., Tian, W., Qiao, K., Zhao, J., Wang, L., Xie, W., Chu, M., Song, T. 2020. Improved chlorine and chromium ion removal from leather processing wastewater by biocharcoal-based capacitive deionization. *Sep. Purif. Technol.*, **233**, 116024.

Fan, X.H., Lin, H., Zhao, J.J., Mao, Y.C., Zhang, J.X., Zhang, H. 2021. Activation of peroxymonosulfate by sewage sludge biochar-based catalyst for efficient removal of bisphenol A: Performance and mechanism. *Sep. Purif. Technol.*, **272**, 8.

Hoslett, J., Ghazal, H., Katsou, E., Jouhara, H. 2021. The removal of tetracycline from water using biochar produced from agricultural discarded material. *Sci. Total Environ.*, **751**, 10.

Hung, C.M., Chen, C.W., Huang, C.P., Lam, S.S., Dong, C.D. 2022. Peroxymonosulfate activation by a metal-free biochar for sulfonamide antibiotic removal in water and associated bacterial community composition. *Bioresour. Technol.*, **343**, 9.

Kumar, A., Saini, K., Bhaskar, T. 2020. Advances in design strategies for preparation of biochar based catalytic system for production of high value chemicals. *Bioresour. Technol.*, **299**, 122564.

Lado, J.J., Zornitta, R.L., Vazquez Rodriguez, I., Malverdi Barcelos, K., Ruotolo, L.A. 2019. Sugarcane biowaste-derived biochars as capacitive deionization electrodes for brackish water desalination and water-softening applications. *ACS Sustaina. Chem. Eng.*, **7**(23), 18992–19004.

Leng, L., Xu, S., Liu, R., Yu, T., Zhuo, X., Leng, S., Xiong, Q., Huang, H. 2020. Nitrogen containing functional groups of biochar: an overview. *Bioresource Technol.*, **298**, 122286.

Libra, J.A., Ro, K.S., Kammann, C., Funke, A., Berge, N.D., Neubauer, Y., Titirici, M.-M., Fühner, C., Bens, O., Kern, J. 2011. Hydrothermal carbonization of biomass residuals: A comparative review of the chemistry, processes and applications of wet and dry pyrolysis. *Biofuels*, **2**(1), 71–106.

Ma, Z., Zhang, Y., Shen, Y., Wang, J., Yang, Y., Zhang, W., Wang, S. 2019. Oxygen migration characteristics during bamboo torrefaction process based on the properties of torrefied solid, gaseous, and liquid products. *Biomass Bioenerg.*, **128**.

Medic, D., Darr, M., Shah, A., Potter, B., Zimmerman, J. 2012. Effects of torrefaction process parameters on biomass feedstock upgrading. *Fuel*, **91**(1), 147–154.

Phanphanich, M., Mani, S. 2011. Impact of torrefaction on the grindability and fuel characteristics of forest biomass. *Bioresour. Technol.*, **102**(2), 1246–1253.

Rouquerol, J., Rouquerol, F., Llewellyn, P., Maurin, G., Sing, K.S. 2013. *Adsorption by powders and porous solids: principles, methodology and applications*. Academic Press.

Wang, R.Z., Huang, D.L., Liu, Y.G., Zhang, C., Lai, C., Wang, X., Zeng, G.M., Gong, X.M., Duan, A., Zhang, Q., Xu, P. 2019. Recent advances in biochar-based catalysts: Properties, applications and mechanisms for pollution remediation. *Chem. Eng. J.*, **371**, 380–403.

Wataniyakul, P., Boonnoun, P., Quitain, A.T., Sasaki, M., Kida, T., Laosiripojana, N., Shotipruk, A. 2018. Preparation of hydrothermal carbon as catalyst support for conversion of biomass to 5-hydroxymethylfurfural. *Catal. Commun.*, **104**, 41–47.

Xu, P., Drewes, J.E., Heil, D., Wang, G. 2008. Treatment of brackish produced water using carbon aerogel-based capacitive deionization technology. *Water Res.*, **42**(10–11), 2605–2617.

Yang, T.T., Xu, Y.M., Huang, Q.Q., Sun, Y.B., Liang, X.F., Wang, L., Qin, X., Zhao, L.J. 2021. Adsorption characteristics and the removal mechanism of two novel Fe-Zn composite modified biochar for Cd(II) in water. *Bioresour. Technol.*, **333**, 9.

10 Valorization of Seafood Processing Discards

Ya-Ting Chen and Shu-Ling Hsieh

CONTENTS

10.1 Introduction	131
10.2 Source of By-Products/Discards	132
10.3 The Environmental Impacts of By-Products/Discards	133
10.4 Bioactive Components of By-Products/Discards	134
10.4.1 Protein	134
10.4.1.1 Protein Hydrolysates	134
10.4.1.2 Collagen and Gelatin	135
10.4.2 Lipid	136
10.4.3 Hydroxyapatite	136
10.4.4 Chitin and Chitosan	136
10.4.5 Astaxanthin	137
10.5 Current Applications of By-Products/Discards	137
10.5.1 Food Industry	137
10.5.2 Pharmaceutical Industry	138
10.6 Extraction Methods to Enhance the Value of By-Products/Discards	139
10.6.1 Biotechnological Approaches	139
10.6.1.1 Enzymatic Extraction	139
10.6.1.2 Fermentative Extraction	140
10.6.2 Physical Approaches	140
10.6.2.1 High Hydrostatic Pressure	140
10.6.2.2 Pulsed Electric Field	140
10.6.2.3 Membrane Separation Technology	141
10.6.2.4 Ultrasound-Assisted Extraction	141
10.6.2.5 Subcritical Water Extraction	141
10.6.2.6 Supercritical Fluid Extraction	141
10.7 Physiological Activity and Application of By-Products/Discards after Mass Conversion/Transformation	141
10.8 Future Prospects and Conclusion	142
References	142

10.1 INTRODUCTION

Seafood is vital for the maintenance of human health due to its high nutritional content of protein, lipids, vitamins, and minerals, making it one of the major staple foods for human consumption; moreover, it is also an important dietary source for many countries. According to the statistics, the global fishery production in 2018 was 179 million tons, of which 156 million tons were for human consumption, which was equivalent to about 20.5 kg per capita per year (FAO, 2020). In addition, the average annual growth rate of edible fish consumption worldwide from 1961 to 2017 was about 3.1% (FAO, 2020). However, with the development of the fisheries economy and fish processing

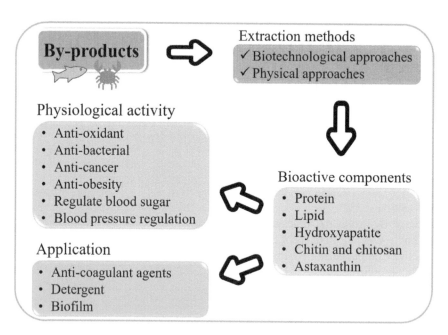

FIGURE 10.1 Sustainable value of seafood processing by-products.

industry, the annual by-products have also increased, up to 22 million tons of marine by-products in 2018. These by-products can be obtained from simple processing due to insufficient storage space on fishing vessels or from processing, cleaning, cooking, cooling, and other procedures. They usually account for 20–70% of raw material depending on the species. Typically, 82% of these by-products are converted to fishmeal, fish oil, fertilizer, and animal feed with low economic value or are buried and discarded. However, these actions may cause serious environmental and economic impacts (FAO, 2020). These by-products contain many nutrients, including proteins, lipids, minerals, vitamins, and chitin. Therefore, much attention has been paid to the means to reuse and develop the by-products into value-added products and further enhance their economic value so as to mitigate their environmental impacts while enhancing the sustainability of fisheries. Recent studies have shown that active substances extracted from by-products, such as protein hydrolysates, collagen, hydroxyapatite, chitin, and astaxanthin, can be used extensively in different domains (Ali et al., 2021). These substances offer more opportunities for the application of the value-added by-products. Even though chemical reagents can be used as extractants in extraction processes to extract a certain amount of active substances at low cost, these extraction processes tend to consume large amounts of organic solvents, which not only cause environmental problems but cannot be used directly as products for human consumption. Therefore, there is still a need to find extraction methods that are non-toxic, solvent selective, environmentally friendly, and able to increase the extraction rate.

In general, these by-products/discards are nutrient rich and can be extracted through different methods to obtain bioactive compounds. They play essential roles in different industries. In this chapter, we present the latest information on marine by-products or processed by-products to recognize the sustainable value of processed marine by-products/discards.

10.2 SOURCE OF BY-PRODUCTS/DISCARDS

Fisheries and aquaculture are not only an important source of animal protein worldwide but also sustain the livelihoods of many. According to FAO statistics, global fishery production reached 179 million tons

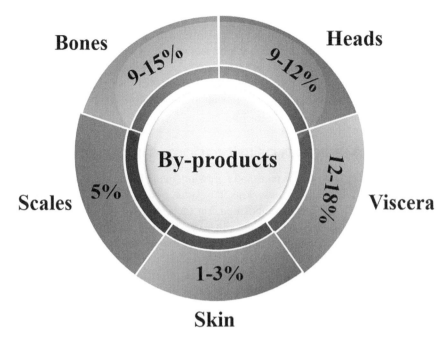

FIGURE 10.2 By-products of fishing processing.

in 2018, of which about 88% (156 million tons) was for direct human consumption and the remaining 12% (22 million tons) was for non-food uses (FAO, 2020). Generally, processed fish remains are termed by-products and are classified as Category 3 by-products according to EU regulations (EC No 1774/2002), meaning animal parts that are suited for human eating but not meant for human consumption. The ratio of edible fish to by-products varies depending on the fishing area, season, and fish size and species. Common by-products from fisheries and aquaculture include fish heads, fins, backbones, gills, belly flaps, roe, skin, and viscera. Moreover, developments in fish processing techniques have increased the number of by-products, which are estimated to account for 20–70% of processed fish (FAO, 2020). By-products of fish processing include heads (9–12% of total fish weight), skin (1–3%), viscera (12–18%), bones (9–15%), and scales (about 5%) (Figure 10.2) (FAO, 2020). Finfish by-products include viscera, heads, skeletal frames, skin, and scales, which usually account for 25–50% of the raw material (Sasidharan and Venugopal, 2020). By-products that may be produced during the fish canning process include skin, muscle, dark flesh, head, and bone, which account for up to 70%. Large fish may produce up to 60% of skeletal frame by-products during filleting. The by-products of crustacean processing account for about 50% of shellfish, which includes cephalothorax, carapace, tail, and shell. The by-products of mussel include shell, meat, extracellular fluid, and byssus threads (Sasidharan and Venugopal, 2020).

10.3 THE ENVIRONMENTAL IMPACTS OF BY-PRODUCTS/DISCARDS

The process generates a large number of by-products as well as a large amount of wastewater from cleaning, cooling, cooking, and curing. Studies have shown that 10–40 m^3 of water is used per ton of seafood after processing. Furthermore, when fishing, fishermen will usually handle the catch in a simple procedure such as removing the offal, head, or skeleton, so as to save space onboard. However, because of the limited availability of refrigeration or freezing facilities for storing economically valuable products on fishing vessels, these unwanted by-products are dumped directly into the ocean or buried. However, these by-products and wastewater can be harmful to the environment.

The anaerobic decomposition of by-products in landfills produces gases such as ammonia, methane, and hydrogen sulfide, which are harmful to the environment. Additionally, the by-products not only decay quickly due to their high protein and nitrogenous substance content but may also reduce the oxygen content of the seafloor, which in turn suffocates marine organisms or introduces diseases, thereby destroying the seafloor ecosystem. The wastewater discharged during fish processing may cause environmental hazards such as eutrophication, habitat destruction, or disease outbreaks because it is not well processed (Venugopal, 2021).

10.4 BIOACTIVE COMPONENTS OF BY-PRODUCTS/DISCARDS

In this era where the fishery economy is constantly growing, many by-products are produced during catching and processing, such as fish heads, skins, fins, offal, bones, and sometimes muscle, which may constitute 10–90% of a seafood's composition. Yet the actual proportion depends on the type and function of seafood. These by-products are often discarded or turned into products with low economic value, which pollutes the environment or even damages the entire marine ecosystem. However, these by-products contain many nutrients, including proteins, lipids, minerals, vitamins, and chitin. The proper utilization of these by-products not only increases their economic value but may also reduce the potential negative environmental impacts.

10.4.1 Protein

Protein is a biological macromolecule necessary for the physiological functioning of living organisms and also one of the six essential nutrients for the human body. These inedible by-products, which are produced through simple handling or processing during the fishing process, are rich in proteins. The bioactive components extracted are not only used as animal feed supplements but also provide the amino acids that are essential to human nutrition. The proteins in marine by-products can be divided into protein hydrolysate, collagen, and gelatin through different extractions, each of which is described in the following.

10.4.1.1 Protein Hydrolysates

Protein hydrolysates are small molecular peptides or amino acids with active functions that are broken down by enzymes or microorganisms through acid-base reactions. Protein hydrolysates are used as ingredients in the food and pharmaceutical industries to improve human health and prevent diseases. It has been proven that protein hydrolysates from various food sources possess a variety of bioactive functions in addition to their nutritional properties (Sila and Bougatef, 2016). Fish proteins contain many nutrients, including essential amino acids for growth, as well as polyunsaturated fatty acids that confer health benefits. Fish proteins also have a higher nutritional value than animal proteins because they contain more essential amino acids such as valine and lysine (López-Pedrouso et al., 2020). In general, raw fish flesh contains about 17–22% (w/w) of crude protein, while crustaceans and mollusks contain 7–23% (w/w). It has also been shown that the crude protein content of fish by-products ranges from 8% to 35%, which indicates that the by-products have high reuse value (Sila and Bougatef, 2016). Fish meat is mainly composed of myofibrillar and sarcoplasmic proteins. Myofibrillar is mainly composed of enzymes related to energy production, while sarcoplasmic is a structural protein that contains actin and myosin. Since hydrolyzed fish proteins offer better digestibility and rapid adsorption properties, as well as bioactive functions for human health, fish proteins have become the main method for many researchers to investigate whether the hydrolysis of by-products can produce bioactive compounds.

Currently, protein hydrolysis can be carried out by enzymatic, microbial, or chemical means (López-Pedrouso et al., 2020). The softness conditions of enzymes can enhance the functional and biological activity of the hydrolyzed products compared to chemical methods. Therefore, proteins are often hydrolyzed using enzymes. The most commonly used enzymes in enzymatic hydrolysis are alcalase, papain, trypsin, flavorzyme, protamex, protease, orientase, neutrase, thermolysin, pancreatin, and validase. Fish by-products, including frames, muscle, viscera, heads, skins, skeletons, and scales, that are hydrolyzed by enzymes may produce protein hydrolysate with antioxidant activity. By-products of carcasses, trimmings, and viscera have antioxidant activity and anti-microbial (anti-bacterial) activity post-hydrolysis. By-products of heads and viscera exhibit antioxidant activity and anti-hypertensive activity after enzymatic hydrolysis. Viscera by-products can produce protein hydrolysate with antioxidant activity and anti-inflammatory activity after enzymatic hydrolysis. Shells, by-product of shellfish, are hydrolyzed by enzymes to produce protein hydrolysates with anti-hypertensive activity.

10.4.1.2 Collagen and Gelatin

Collagen is a kind of protein that is the main fibrous and structural protein of the animal extracellular matrix and contributes to the physiological function of tissues such as bones, skin, tendons, and cartilage (Al Khawli et al., 2020). Collagen is a helical macromolecule mainly composed of three α-chains. Each α-chain is formed by specific repeating elements of glycine-proline-hydroxyproline triplet. So far, at least 28 different types of collagens have been identified, with type I collagen being the most common form in animals, mainly found in the tissues of bones, skin, and tendons (Jafari et al., 2020). Gelatin is an irreversible protein molecule formed by thermal and kinetic denaturation of collagen. Its composition is similar to collagen and also shares some similar properties with collagen (Al Khawli et al., 2020). Collagen and gelatin are commonly used in food, pharmaceutical, biomedical, and medical applications due to their biodegradability and good biocompatibility as well as technical applicability. Most collagen and gelatin available on the market is made of pig and cattle hides, which account for about 46% of world production, and only about 1% comes from marine sources. However, consumers are concerned about these products due to risk factors that are harmful to human health, such as hoof-and-mouth disease (*Aphtae epizooticae*) from pigs and bovine spongiform encephalopathy from cattle (Ferraro et al., 2010). In addition, some countries are not able to accept collagen and gelatin-related products made from pigs and cattle due to religious restrictions (Bruno et al., 2019). For this reason, collagen made of marine sources seems to be a good alternative. In particular, collagen from marine by-products, including skin, bone, scales, swim-bladder, head, and cartilage, has a similar amino acid composition and biocompatibility compared to porcine or bovine collagen but with more bioavailability and has up to 1.5 times higher absorption efficiency due to its small particle size and low molecular weight (Jafari et al., 2020). Therefore, marine by-products are a safe and reliable source of collagen.

The current methods used to isolate collagen from marine by-products include acid-solubilized collagen, extraction of pepsin-solubilized collagen, deep eutectic solvent, and supercritical fluid extraction. For collagen, fish by-products, including skin, bone, and scales, are hydrolyzed by enzymes to produce collagen with antioxidant activity. Hepatopancreas and viscera are hydrolyzed by enzymes to produce collagen with antioxidant activity and anti-hypertensive activity. The collagen isolated from by-product of scales through enzymatic hydrolysis not only has antioxidant activity but also anti-inflammatory activity. The collagen isolated from skin through enzymatic hydrolysis has both antioxidant activity and anti-glycation activity. In addition, the collagen isolated from skin through enzymatic hydrolysis can enhance bone strength and promote wound healing. In studies on gelatin, enzymatic hydrolysis of skeleton can increase gelatin production. The enzymatic hydrolysis of skin and scale produces gelatin with antioxidant activity. In addition, the collagen isolated from skin through enzymatic hydrolysis has antioxidant and anti-hypertensive activity.

10.4.2 LIPID

Lipid is not only an important source of energy for living organisms but is also a major component of cell membranes. It is also involved in many cellular signaling pathways (Ferraro et al., 2010). Although polyunsaturated fatty acids in lipids are among the essential nutrients for human health, they cannot be synthesized by the human body and must be procured from food. Among the polyunsaturated fatty acids, omega-3 is of particular importance (López-Pedrouso et al., 2020). Omega-3 is a long-chain polyunsaturated fatty acid with a chain length of 18, 20, or 22 carbon atoms. It mainly comes from seafood or other food products. Omega-3 primarily contains alpha-linolenic acid (ALA), eicosapentaenoic acid (EPA), and docosahexaenoic acid (DHA), which are commonly used as functional food components, drugs, and nutritional supplements because of their ability to prevent cardiovascular disease, lower blood pressure, inhibit platelet aggregation, and reduce inflammatory responses. Marine fish with high fat content, such as salmon, sardine, herring, and mackerel, are the main sources of omega-3 (Al Khawli et al., 2020).

The current methods for extracting omega-3 from marine organisms are solvent-based extraction, supercritical extraction, enzymatic extraction, microwave-assisted extraction, and ultrasonic extraction. Fish by-products, including gills, gut, skin, meat, bones, heads, frame, fins, tails, viscera, and the viscera of cephalopods, can produce fish oil with good quality and high omega-3 content through solvent extraction, enzyme extraction, high hydrostatic pressure extraction, subcritical water extraction, or cooking chambers.

10.4.3 HYDROXYAPATITE

Hydroxyapatite [$Ca_{10}(PO_4)_6(OH)_2$, HAp] belongs to the family of calcium phosphate, and its composition is very similar to that of bone mineral (Piccirillo et al., 2015). Because HAp can be combined with bone without causing local or systemic toxicity, foreign body response, or even inflammation, and it is also highly biocompatible, it is widely used in orthopedic implants, dental materials, and other biomedicine applications (Piccirillo et al., 2015). HAp can be chemically synthesized or isolated from natural resources. The methods used to synthesize HAp are solid-state synthesis, hydrothermal treatment, chemical precipitation, radio frequency thermal plasma, and polymer-assisted methods. However, the synthesis is a challenging process because toxic intermediates can potentially be formed in addition to the high cost and complexity of the synthesis methods (Ideia et al., 2020). HAp can be synthesized from natural resources through calcination, alkaline hydrolysis, enzymatic hydrolysis, and ionic liquid pretreatment (Ideia et al., 2020). Extraction of HAp from marine by-products such as fish bones and scales is not only inexpensive and simple but also more compatible than synthetic methods.

Fish by-products, including jaw, scales, bones, heads, frames, and shells from shellfish and crustaceans, can be used to produce HAp through calcination, enzymatic hydrolysis, wet-chemistry, wet precipitation, precipitation, and hydrothermal methods. HAp can be used in bioimaging and skeletal correlation and contribute to bone metabolism and cell adhesion.

10.4.4 CHITIN AND CHITOSAN

Chitin is a natural polymer which is insoluble in water and organic solvents. It has a white or yellowish color, is odorless and tasteless and widely found in nature, and is the second most abundant polysaccharide after cellulose (Al Khawli et al., 2020; Santos et al., 2020). Chitin can be extracted from the shells of invertebrates such as shellfish and crustaceans. It can also be further converted to chitosan through deacetylation (Santos et al., 2020). Chitin and chitosan are widely used in the food, cosmetic, pharmaceutical, agricultural, paper, and environmental industries, as they are non-toxic polymers that are water insoluble, biocompatible, biodegradable, and capable of being regenerated and forming gels. Crustacean shells contain 13–42% of chitin. This suggests that using marine by-products to extract chitin is not only natural, low cost, and able to increase the value of by-products but can also reduce environmental pollution.

Currently, both chemical and biological methods are used to extract chitin. Chitin can be extracted from crustacean shells and other by-products through chemistry, enzymatic hydrolysis, or fermentation and further deacetylated to chitosan for medical, pharmaceutical, and biotechnological applications. Furthermore, chitin can be extracted from crustacean shells through chemistry, microwave methods, or enzymatic hydrolysis and further deacetylated to chitosan. Chitin and chitosan both have anti-microbial activity and antioxidant activity, anti-inflammatory activity, and anticancer activity and can potentially treat hypercholesterolemia.

10.4.5 Astaxanthin

Astaxanthin (3,3-dihydroxy-β,β-carotene-4,4-dione) is a member of the carotenoids family. A natural pigment with both hydrophilic and lipophilic properties, it is widely present in crustacean shells, plants, and microorganisms, and accounts for 74–98% of the total pigments in crustacean shells (Bruno et al., 2019; Zhao et al., 2019). Astaxanthin has a wide range of biological activities, with its antioxidant capacity being the strongest, at up to 10 times higher than other carotenoids and up to 500 times higher than vitamin E. In addition, astaxanthin is widely used in the food, cosmetics, feed, and pharmaceutical industries. However, 95% of astaxanthin around the world is synthesized by using petrochemical resources and cannot be used in human food or additives due to food safety, environmental, and sustainability concerns. Therefore, marine by-products could be a good alternative.

Currently, solvent extraction, oil stripping, enzymatic hydrolysis, supercritical fluid extraction, magnetic-field-assisted extraction, ultrasound assisted extraction, high-pressure processing methods, soxhlet extraction, and microbial fermentation are commonly used to extract astaxanthin (Zhao et al., 2019). Shells, heads, exoskeletons, and other by-products of crustaceans can be extracted through solvent extraction or enzymatic hydrolysis. In addition, crustacean shells, heads, and other by-products extracted through high pressure extraction, supercritical fluid extraction, solvent extraction, or enzymatic hydrolysis offer antioxidant activity and anticancer effects.

10.5 CURRENT APPLICATIONS OF BY-PRODUCTS/DISCARDS

According to FAO statistics, up to 22 million tons (12%) of the global fishery production in 2018 was for non-food purposes, 20–70% of which were by-products from fishing or processing (FAO, 2020). However, these by-products have beneficial nutrient components for human health, such as proteins, lipids, chitin, and astaxanthin. In the past, by-products were mostly low economic value products such as fish meal, fodder, and fertilizer. In fact, these by-products have potential applications in food, agriculture, medicine, pharmaceuticals, and the environment. Most of them, however, are dominated by the food and pharmaceutical industries; therefore, the following is a description of the current applications of marine by-products in the food and pharmaceutical industries.

10.5.1 Food Industry

Most marine by-product protein recycling processes involve hydrolysis, which not only retains the amino acid content but also enhances the physical and chemical properties of proteins, making them suitable for a wide range of applications in different industries. Protein hydrolysates produced through hydrolysis of marine by-products can potentially be used as food ingredients, as they enhance solubility, emulsification, and oil binding capacity, which are crucial in food formulations (Aspevik et al., 2017). In addition, protein hydrolysates produced through the hydrolysis of by-products can also be isolated as biologically active peptides, including anti-bacterial, antioxidant, anti-hypertensive, anti-proliferative, anti-coagulant, and immunomodulatory effects. Therefore, they can be used in food additives or nutritional supplements.

Collagen extracted from marine by-products such as skin, bone, fish scales, and cartilage is often used as food additives to improve texture, stability, and water holding capacity of food products due to its biodegradability and good biocompatibility. In addition, collagen could also be used in combination with pectin to improve the physical and chemical properties of traditional thickening agents or gel. Moreover, the ability of collagen to gelatinize allows it to be used as sausage coating. A film made of collagen acts as a barrier to water and oxygen and inhibits undesirable biochemical reactions, thus prolonging the shelf life of meat products and reducing aroma and product deterioration.

Gelatin is a derivative of collagen and can be extracted from the skin of various fish species, including sea bream, giant catfish, striped catfish, and leather-jacket (Shavandi et al., 2019). Gelatin is widely used in the food industry due to its transparency, low sweetness, good gel forming ability, and ability to undergo thermo-reversible gelation. For instance, the thermo-reversible gelation of gelatin and water can be used in jelly production. In addition, gelatin can also be used as a gelling agent in chewing gum. Gelatin can also be used in ice cream or other desserts, as it prevents the forming of ice crystals in the product structure and provides a good creamy texture when it dissolves in the mouth (Shavandi et al., 2019). Moreover, gelatin can be used in edible films, packaging, dairy processing, nutritional supplements, texturizers, foaming agents, and colloidal stabilizers for food industry applications.

Fish oil extracted from marine by-products such as gills, skin, viscera, and heads contains high-quality omega-3 fatty acids, which are important in the food industry because they confer health promotion benefits, promote reproductive development, and reduce the risk of developing many diseases. Fish oil is rich in omega-3 fatty acids, including eicosapentaenoic acid (EPA) and docosahexaenoic acid (DHA), and therefore could be used in the production of margarine or added directly to milk powder and other foods (Senevirathne and Kim, 2012).

Chitin is mostly extracted from the shells of crustaceans and shellfish. With properties such as being biodegradable, non-toxic, anti-bacterial, and gel-forming and having protein binding affinity, chitin is applied in food and other industrial applications. Chitin can be used in the clarification of fruit juice, milk processing, food packaging, or coating materials to extend the shelf life of foods (Senevirathne and Kim, 2012).

Chitosan is derived from the deacetylation of chitin, and it is commonly used in the food industry as a dietary food additive and natural preservative for foods such as meat, fruits, and vegetables and as a compound for preserving and clarifying beverages (Bakshi et al., 2020). Chitosan is also used as a dietary supplement for obese individuals because it cannot be absorbed by the human body and can be eliminated by binding itself to fat and bile acids. In addition, chitosan is biodegradable and acts as a gas and aroma barrier, making it environmentally friendly for food packaging materials.

Astaxanthin is a characteristic pigment of crustaceans that cannot be synthesized by animals. It is produced through the food chain when it enters an animal's body. Because of its antioxidant activity, astaxanthin is widely used as a nutritional supplement.

10.5.2 Pharmaceutical Industry

The matrix proteins extracted from hydrolyzed shell proteins are used as ingredients in cosmetics and as hair and skin conditioners (Olsen et al., 2014). Collagen extracted from marine by-products is widely used in the cosmetics, medical, and pharmaceutical industries. Due to its water holding capacity, tissue repair, and biocompatibility, collagen is used in beauty or cosmetic products to maintain water holding capacity in the skin and promote tissue repair. In the medical and pharmaceutical industries, collagen is used as a drug carrier and gene delivery agent for bone and cartilage formation, supplement for treating weak nails and nourishing scalp hair, wound dresser, and vitreous implant.

Gelatin has been widely used in the cosmetics, medical, and pharmaceutical industries due to its good biodegradability, biocompatibility, and cell adhesion properties. Gelatin has been used in the development of cosmetic products because it promotes skin repair, tissue regeneration, and mitigation of UV-induced skin damage. Gelatin is used in the medical and pharmaceutical industries as a drug carrier, drug for cartilage and bone formation, and in tissue engineering and implants. In addition, gelatin is also commonly used in the manufacture of capsules, ointments, and emulsions.

Hydroxyapatite is mainly extracted from marine by-products such as fish bones and scales and is commonly used in commercial fertilizers. However, because it is less harmful to humans and the environment, as well as the increasing global demand for phosphorus/phosphate, hydroxyapatite of natural origin is has gained popularity. In addition, bones are mainly composed of type I collagen in the organic phase and calcium phosphate in the inorganic phase. Hydroxyapatite, in particular, is extracted from marine by-products and has a composition similar to that of bone mineral, allowing it to bind to bone easily, and is highly biocompatible. For this reason, hydroxyapatite can be used in biomedical applications. Due to its osteoconductive properties, hydroxyapatite facilitates the growth of osteoblasts and new bone formation, and it is widely used as an alternative source of chemically synthesized hydroxyapatite for bone tissue engineering such as orthopedic implants and dental materials, as well as biomedical materials.

Besides the food industry, chitin is also widely used in medicine and pharmaceutical industry as an excipient, wound dressing material, and drug carrier. In addition, chitin can form composites of hydroxyapatite and chitosan, which can be used as bone filling materials for bone tissue regeneration therapy.

Chitosan is a natural pseudocationic polymer that forms a moisturizing elastic film on the skin surface, protecting the lips from dryness and softening them. It also has deodorizing, anti-bacterial, and higher scent adhesion properties, making it useful for cosmetic applications such as moisturizers and hair and skin care (Bakshi et al., 2020). In medicine, chitosan has been proved to be a safe drug formulation with antacid and anti-ulcer properties. It can also prevent stomach irritation, making it widely used as an emulsifier in pharmaceuticals. In addition, chitosan has been widely used as a carrier for drug delivery, including delivery of proteins/peptides, anti-inflammatory drugs, anti-biotics, and vaccines, due to its non-toxic and bioresorbable properties.

10.6 EXTRACTION METHODS TO ENHANCE THE VALUE OF BY-PRODUCTS/DISCARDS

The most commonly used extraction method for marine by-products is solvent extraction, which can extract different active substances depending on the nature of the solvent. However, using solvent extraction not only consumes large amounts of organic solvents that impact the environment but may also reduce the yield, quality, and value of active substances. Most importantly, these solvents pose safety concerns for human consumption. Therefore, there is a need to develop extraction methods that have high extraction rates, short processing times, and low toxicity; involve no or reduced solvent usage; are environmentally friendly; and maximize the conversion of active substances from marine by-products so as to enhance their value and reduce environmental problems. The extraction methods introduced as follows offer high extraction rates, low toxicity, short processing times, solvent selectivity, and eco-friendliness and can extract high-value active substances from marine by-products.

10.6.1 Biotechnological Approaches

10.6.1.1 Enzymatic Extraction

Enzyme hydrolysis is mainly used to extract active substances from food, food by-products, industrial by-products, or marine by-products by breaking the cell wall and releasing bioactive substances. Compared with conventional solvent extraction, enzymes are highly specific, environmentally friendly, and safe and have specific and regioselective catalytic reactions in aqueous forms. Therefore, they can be used as an alternative to conventional solvent extraction (Bruno et al., 2019; Ali et al., 2021). In addition, enzymes used in enzyme hydrolysis, such as alcalase, flavorzyme, protemax, pepsin, papain, devolvase, trypsin, and pancreatin, are usually isolated from microorganisms, animals, and plants, making them more environmentally friendly than conventional methods. They are therefore considered an environmentally friendly extraction method. Currently, enzymatic hydrolysis is commonly used to extract active substances such as fish oil and proteins from marine by-products.

10.6.1.2 Fermentative Extraction

Fermentation is a natural decomposition process that uses microorganisms to transform complex organic substances into simple compounds, which leads to the production of fermentation by-products or hydrolysis of target compounds (Marti-Quijal et al., 2020). Fermentation is widely used in food preservation not only to increase shelf life, flavor, texture, and aroma but also to produce bioactive substances with functional properties. Because of the short shelf life of fresh fish, fermentation is also used in many countries around the world to extend the shelf life of fish. Common fermented fish products include fish sauce and fish paste. Therefore, fermentation is a safe and environmentally friendly technology. A variety of compounds, such as bioactive peptides or aromatic compounds, can be obtained through fermentation. Currently, fermentation is widely used in marine by-products to extract active ingredients such as proteins, lipids, chitin, chitosan, and astaxanthin.

10.6.2 Physical Approaches

10.6.2.1 High Hydrostatic Pressure

High hydrostatic pressure is a non-thermal processing technology that has been used to reduce microorganisms and inactivate enzymes in seafood, dairy products, and vegetables for food preservation and sterilization purposes (Bruno et al., 2019; Ali et al., 2021). In high hydrostatic pressure processing, the ability of cell membranes to deprotonate under high pressure and break salt bridges and weak bonds is leveraged to increase membrane permeability and facilitate the release of active substances. High hydrostatic pressure processing is not only an innovative and environmentally friendly extraction technique but also provides better extraction results than conventional methods and other techniques. Currently, high hydrostatic pressure processing has been used to extract active substances such as proteins, astaxanthin, and lipids from marine by-products.

10.6.2.2 Pulsed Electric Field

In pulsed electric field processing, a certain level of electric field intensity is applied to biological cells, which causes damage to the cell membrane structure and creates pores, resulting in increased permeability, which in turn facilitates the release of active substances. Pulsed electric field processing has the potential to be used as an alternative to conventional extraction methods because of its short processing time, low cost, fast response time, zero usage of harmful solvents, and higher extraction rates compared to conventional thermal extraction-based techniques. Currently, pulsed electric field processing is used to extract substances with antioxidant activity from fish by-products. However, pulsed electric field processing may damage samples due to electrode corrosion and electrode migration during extraction. Therefore, their applicability for marine by-products warrants further investigation.

Valorization of Seafood Processing Discards

10.6.2.3 Membrane Separation Technology

Membrane separation technology is an extraction technique that uses membranes as a selective or semi-permeable barrier to separate molecules based on their size and permeability. Compared to conventional extraction techniques, membrane separation technology does not require any reagents or destructive compounds and is environmentally sustainable. It therefore could potentially replace conventional extraction techniques (Ali et al., 2021). Currently, membrane separation technology is used in the nutraceutical and cosmetic industries, as well as to extract nutritive or active substances (including proteins, fatty acids, enzymes, and biomaterials) from marine by-products. However, the effectiveness of membrane separation technology may be affected by microorganisms as well as inorganic or organic compounds, resulting in different concentrations of active substances. Hence, there is a need to further study the applicability of membrane separation technology.

10.6.2.4 Ultrasound-Assisted Extraction

Ultrasound technology has been applied in the food industry for the main purpose of eliminating microorganisms. Ultrasound technology uses sound waves to produce cavitational effects that compress and break cell walls and accelerate heat and mass transfer in cells, resulting in rapid release of active substances. Ultrasound can potentially be used as an alternative to traditional extraction techniques due to its ability to improve food properties and eliminate microorganisms, high extraction rate, low solvent and emulsifier consumption, safety, and reusability. Ultrasound-assisted extraction has been widely used to extract active substances such as collagen, chitin, and proteins from marine by-products (Ali et al., 2021). However, the use of higher-intensity sound waves and longer extraction time can decrease the functionality and quality of active substances. Thus, the intensity of sound waves used during extraction as well as the extraction time are two important factors to be considered.

10.6.2.5 Subcritical Water Extraction

Subcritical water extraction is an extraction method that uses water as a solvent and maintains its liquid form under high temperatures and certain pressure ranges (Plaza et al., 2010). Subcritical water extraction is an environmentally friendly method that not only improves extraction speed and reduces solvent use but also increases the extraction rate and yield. Therefore, it has the potential to be an alternative to conventional extraction techniques. Currently, subcritical water extraction has been widely used for the extraction of active ingredients such as proteins, lipids, and chitin in marine by-products.

10.6.2.6 Supercritical Fluid Extraction

Supercritical fluid extraction is a method for separating solutes from liquid or solid substances using a solvent at or near the critical temperature and pressure (Bruno et al., 2019). Since the extraction solvent chosen for supercritical fluid extraction is non-toxic; has low critical constant; and is chemically inert, low cost, and easily recyclable, it is considered one of the best alternatives to conventional solvent extraction. Currently, supercritical fluid extraction is mostly used for lipid extraction in marine by-products.

10.7 PHYSIOLOGICAL ACTIVITY AND APPLICATION OF BY-PRODUCTS/DISCARDS AFTER MASS CONVERSION/TRANSFORMATION

The by-products of fishing and fishery processing have become a major problem for the environment and processing plants. Marine by-products are not only enriched with nutritious active substances but can also be extracted through different methods to increase the yield of active substances and even produce functional active peptides or substances, including antioxidant, anti-bacterial, anti-cancer, and anti-obesity action and regulation of blood sugar. In addition, many active substances have been extracted from marine by-products, and their potential application as food or for other industrial purposes has been explored, including anticoagulants agents, detergents, and biofilms.

10.8 FUTURE PROSPECTS AND CONCLUSION

Current developments in the fishery economy, along with the growth of fishery production, have increased the production of marine by-products in both fishing and processing. These by-products are rich in nutrients and nitrogenous substances, which not only cause a negative impact on the environment but may also lead to destroying the marine ecosystem and even severely impact the sustainability of fisheries developments. Many studies have shown that extracting and isolating bioactive substances from the by-products produced by different marine species not only have physiological activity but can also be applied in food, cosmetics, medicine, and pharmaceutics. Although various extraction methods with high extraction rates, low toxicity, solvent selectivity, and eco-friendliness have been proposed as potential alternatives to conventional methods of extracting active substances from marine by-products, these environmentally friendly extraction methods still have many limitations. These include possible sample damage during extraction, which may reduce the extraction rate, or interference by microorganisms or organic/inorganic compounds that may affect the concentration of active substances, as well as other potential problems related to cost, processing issues, and consumer acceptance. Therefore, more research is needed to ascertain the stability and applicability of these extraction methods. Furthermore, the potential for product development and functional application of these active substances present in the by-products needs to be thoroughly investigated.

REFERENCES

Ali, A., Wei, S., Liu, Z., Fan, X., Sun, Q., Xia, Q., Liu, S., Hao, J., Deng, C., 2021. Non-thermal processing technologies for the recovery of bioactive compounds from marine by-products. LWT 147, 111549. https://doi.org/10.1016/j.lwt.2021.111549

Al Khawli, F., Martí-Quijal, F.J., Ferrer, E., Ruiz, M.J., Berrada, H., Gavahian, M., Barba, F.J., de la Fuente, B., 2020. Aquaculture and its by-products as a source of nutrients and bioactive compounds. Food Nutr. Res. 92, 1–33. https://doi.org/10.1016/bs.afnr.2020.01.001

Aspevik, T., Oterhals, Å., Rønning, S.B., Altintzoglou, T., Wubshet, S.G., Gildberg, A., Afseth, N.K., Whitaker, R.D., Lindberg, D., 2017. Valorization of proteins from co-and by-products from the fish and meat industry. Top. Curr. Chem. (Cham). 375, 53. https://doi.org/10.1007/s41061-017-0143-6

Bakshi, P.S., Selvakumar, D., Kadirvelu, K., Kumar, N.S., 2020. Chitosan as an environment friendly biomaterial- a review on recent modifications and applications. Int. J. Biol. Macromol. 150, 1072–1083. https://doi.org/10.1016/j.ijbiomac.2019.10.113

Bruno, S.F., Ekorong, F.J.A.A., Karkal, S.S., Cathrine, M.S.B., Kudre, T.G., 2019. Green and innovative techniques for recovery of valuable compounds from seafood by-products and discards: a review. Trends Food Sci. Technol. 85, 10–22. https://doi.org/10.3390/md15050131

FAO, 2020. *The State of World Fisheries and Aquaculture 2020*. Sustainability in Action, Rome. https://doi.org/10.4060/ca9229en

Ferraro, V., Cruz, I.B., Jorge, R.F., Malcata, F.X., Pintado, M.E., Castro, P.M., 2010. Valorisation of natural extracts from marine source focused on marine by-products: a review. Food Res. Int. 43, 2221–2233. https://doi.org/10.1016/j.foodres.2010.07.034

Ideia, P., Pinto, J., Ferreira, R., Figueiredo, L., Spínola, V., Castilho, P.C., 2020. Fish processing industry residues: a review of valuable products extraction and characterization methods. Waste Biomass Valorization 11, 3223–3246. https://doi.org/10.1007/s12649-019-00739-1

Jafari, H., Lista, A., Siekapen, M.M., Ghaffari-Bohlouli, P., Nie, L., Alimoradi, H., Shavandi, A., 2020. Fish collagen: extraction, characterization, and applications for biomaterials engineering. Polymers 12, 2230–2236. https://doi.org/10.3390/polym12102230

López-Pedrouso, M., Lorenzo, J.M., Cantalapiedra, J., Zapata, C., Franco, J.M., Franco, D., 2020. Aquaculture and by-products: challenges and opportunities in the use of alternative protein sources and bioactive compounds. Adv. Food Nutr. Res. 92, 127–185. https://doi.org/10.1016/bs.afnr.2019.11.001

Marti-Quijal, F.J., Remize, F., Meca, G., Ferrer, E., Ruiz, M.J., Barba, F.J., 2020. Fermentation in fish and by-products processing: an overview of current research and future prospects. Curr. Opin. Food Sci. 31, 9–16. https://doi.org/10.1016/j.cofs.2019.08.001

Olsen, R.L., Toppe, J., Karunasagar, I., 2014. Challenges and realistic opportunities in the use of by-products from processing of fish and shellfish. Trends Food Sci. Technol. 36, 144–151. https://doi.org/10.1016/j.tifs.2014.01.007

Piccirillo, C., Pullar, R.C., Costa, E., Santos-Silva, A., Pintado, M.M.E., Castro, P.M., 2015. Hydroxyapatite-based materials of marine origin: a bioactivity and sintering study. Mater. Sci. Eng. C. 51, 309–315. https://doi.org/10.1016/j.msec.2015.03.020

Plaza, M., Amigo-Benavent, M., del Castillo, M.D., Ibáñez, E., Herrero, M., 2010. Neoformation of antioxidants in glycation model systems treated under subcritical water extraction conditions. Food Res. Int. 43, 1123–1129. https://doi.org/10.1016/j.foodres.2010.02.005

Santos, V.P., Marques, N.S., Maia, P.C., Lima, M.A.B.D., Franco, L.D.O., Campos-Takaki, G.M.D., 2020. Seafood waste as attractive source of chitin and chitosan production and their applications. Int. J. Mol. Sci. 21, 4290. https://doi.org/10.3390/ijms21124290

Sasidharan, A., Venugopal, V., 2020. Proteins and co-products from seafood processing discards: their recovery, functional properties and applications. Waste Biomass Valorization 11, 5647–5663. https://doi.org/10.1007/s12649-019-00812-9

Senevirathne, M., Kim, S.K., 2012. Utilization of seafood processing by-products: medicinal applications. Food Nutr. Res. 65, 495–512. https://doi.org/10.1016/B978-0-12-416003-3.00032-9

Shavandi, A., Hou, Y., Carne, A., McConnell, M., Bekhit, A.E.D.A., 2019. Marine waste utilization as a source of functional and health compounds. Food Nutr. Res. 87, 187–254. https://doi.org/10.1016/bs.afnr.2018.08.001

Sila, A., Bougatef, A., 2016. Antioxidant peptides from marine by-products: isolation, identification and application in food systems. A review. J. Funct. Foods 21, 10–26. https://doi.org/10.1016/j.jff.2015.11.007

Venugopal, V., 2021. Valorization of seafood processing discards: bioconversion and bio-refinery approaches. Sustain. Food Syst. 5, 611835. https://doi.org/10.3389/fsufs.2021.611835

Zhao, T., Yan, X., Sun, L., Yang, T., Hu, X., He, Z., Liu, F., Liu, X., 2019. Research progress on extraction, biological activities and delivery systems of natural astaxanthin. Trends Food Sci. Technol. 91, 354–361. https://doi.org/10.1016/j.tifs.2019.07.014

11 Effect of Microplastics on Marine Ecosystems

De-Sing Ding and Anil Kumar Patel

CONTENTS

11.1	Introduction	145
11.2	Sources of Microplastics in the Ocean	145
11.3	The Threat of Microplastics to Aquatic Organisms in the Ocean	147
11.4	What Nutrition Do Microplastics Have and Why Do They Replace the Status of Live Food?	148
11.5	The Impact of Microplastics on Marine Ecosystems	149
11.6	The Threat of Microplastics to Aquaculture	152
11.7	Actions to Protect Marine Ecological Sustainability	153
11.8	Conclusions and Perspectives	154
References		154

11.1 INTRODUCTION

Taiwan is an island country located in the north Pacific Ocean and has abundant marine resources. In Taiwan's southwest, on the Gold Coast in Tainan City, powerful waves crash against the black sand. It has idyllic tropical beaches, rich in biodiversity; many people play and fish here; and it is also an important oyster farming area. But the Gold Coast is often covered in plastic; it is like the beach which is covered with the plastic carpet (Figure 11.1). According to observations, in addition to fishing nets, ropes, buoys, and so on, which are commonly found in marine plastic waste, a large amount of household waste and packaging plastics can also be found, such as plastic bottles, plastic lunch boxes, shoes, plastic straws, and toothbrushes. These plastic wastes are washed up on the beach due to the combined action of ocean currents and local eddies, which makes the originally beautiful beach look like a large garbage dump. This not only affects the appearance but also poses a major threat to marine ecology. The same beach plastic waste hazard occurs every day in oceans around the world. In addition to Taiwan, in Nagasaki, Japan (latitude and longitude 32°43′14.8″N 129°50′47.1″E), there are also traces of plastic waste found along the coast and ports (Figure 11.2). The problem of litter in the ocean has been regarded by scientists as one of the threats to biodiversity (Gall and Thompson, 2015) and may threaten human activities and health. According to scholars, garbage has also accumulated on the beach at Kamilo Beach, which is known as one of the dirtiest beaches in the world (Cressey, 2016). Therefore, marine plastic waste is a global problem. These marine plastics will be transported to various places with ocean currents, which will make the ocean like a garbage dump and cause serious harm to the marine ecology, marine environment, and health of marine life. It has become a global concern for pollution due to its prevalence and persistence in the aquatic environment and its potential risk to ecosystem health. According to the research, about 75% of the garbage found on the coastline of the world is plastic products. Therefore, most of the waste in the ocean is plastic waste (Derraik, 2002). We will explore, one by one, what the impact of these marine plastics is on marine biological systems.

11.2 SOURCES OF MICROPLASTICS IN THE OCEAN

Plastics are widely used in many aspects of life due to their excellent characteristics such as light weight, strong plasticity, strong flexibility, thermal insulation and electrical insulation, corrosion

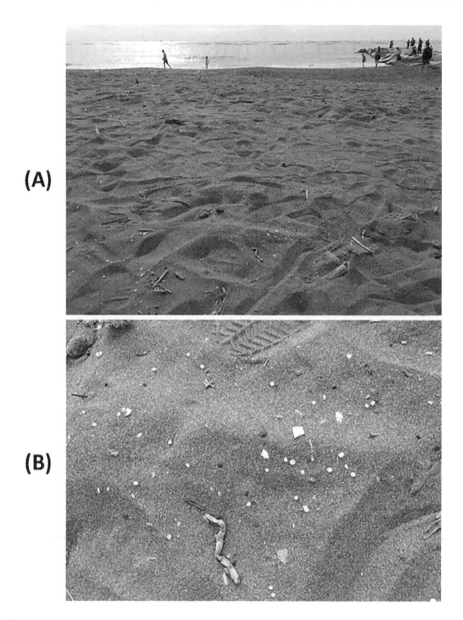

FIGURE 11.1 Microplastics found on gold coast in Tainan city (latitude and longitude 22°55'55.7"N 120°10'33.8"E). Shooting time 2021/10/30. A: Microplastics are scattered on the beach; B: microplastics are degraded to produce different particle types.

resistance, and low cost. After the mass production of plastic products, and due to improper management and disposal, a large amount of plastic waste enters the environment through various channels, causing serious environmental pollution problems (Geyer et al., 2017). Plastic has been manufactured in large quantities since industrialization, and more than 40% of it is used as disposable packaging. Discarded plastic eventually becomes plastic waste and accumulates in soil and water resources (Andrady, 2011). After these plastic wastes enter the natural environment, they are broken into small pieces due to photo-oxidation, thermal reactions, microorganisms, and mechanical forces. The materials of these plastic particles are nylon, polyethylene, polypropylene, polystyrene, and acrylic (Cole et al., 2013). In addition, plastic waste on land will slowly decompose under the

FIGURE 11.2 Plastic waste found on the coast of Nagasaki, Japan, and ports (latitude and longitude 32°43′14.8″N 129°50′47.1″E). Shooting time 2022/04/30. A: Plastic garbage scattered on the coast; B: plastic garbage floating on the sea of the port.

action of physics, chemistry, and biology to produce many smaller microplastics. Microplastics are defined as plastic fragments less than 5 mm in diameter (Thompson et al., 2004). When solid plastic is composed of less than 5 mm particles, it exhibits the characteristics small size, and light weight while carrying the same volume. Moreover, due to small size, it cannot be collected and recycled, and it will drift into the sea with the confluence of rivers.

In addition, plastic particles are often added as an ingredient in products. In daily cleaning products, such as toothpaste, shower gel, and facial cleanser, plastic particles are sometimes added to improve the scrubbing effect. However, at present, some products have been gradually controlled internationally, hoping to effectively prevent plastic particles from directly flowing into the ocean through the sewage system. If these microplastics flow into the sewer system, they will end up drifting in the ocean or deposited in the abyss (Woodall et al., 2014). The Great Pacific Garbage Patch is one of the well-known giant garbage accumulation areas, located between California and Hawaii, and 94% of it is microplastics (Lebreton et al., 2018).

11.3 THE THREAT OF MICROPLASTICS TO AQUATIC ORGANISMS IN THE OCEAN

According to research, plastics are found in the seas from the Arctic to the Antarctic. A small amount of human-made garbage, such as iron or corrodible garbage, will disappear due to decomposition and rot or rust, but plastics can last for many years without disintegrating. Microplastics pose a major threat to marine life and coastlines (Anderson et al., 2016). These tiny microplastics are more easily transported to water environments such as coastal and port areas with natural forces such as wind, rain, and currents. Through these processes, the surface of microplastics will be highly oxidized, which is conducive to the growth of biofilm and the mixing of marine microorganisms and biological feces, thereby increasing the density of microplastics and making them sink into the sediment (Castro et al., 2020). After these microplastics invade the marine ecosystem, they not only pose a threat to the marine environment but also cause physiological harm to marine organisms. Compared with large and small plastic waste, microplastics have the most direct impact on marine organisms. Microplastics are easily eaten by various marine organisms, or microplastics are adsorbed through the gill tissue of fish during respiration, which affects the gas replacement function. From a macro perspective, microplastics can indeed cause different levels of impact on invertebrates and vertebrates in the ocean (Lusher et al., 2013). Previous studies have found that corals, shrimps, and others will eat microplastics by mistake, resulting in microplastic residues in the body, which means that these marine organisms cannot distinguish between live food and microplastics and mistakenly swallow microplastics (Devriese et al., 2015; Ding et al., 2019; Hsieh et al., 2021). In addition, some studies have pointed out that after scleractinian coral *Stylophora pistillata* and *Goniopora columna* are stimulated by microplastics, it will lead to the proliferation of zooxanthellae in the initial stage, resulting in changes in the original balanced symbiosis between coral and zooxanthellae, and a large amount of mucus is produced. Aggregated and precipitated microplastics on coral surfaces and water are largely due to adhesion to the secreted mucus (Chen et al., 2022; Lanctôt et al., 2020). It is not yet known why microplastics can cause the growth of zooxanthellae in corals, but it is known that after long-term observation, these zooxanthellae will decrease over time, and finally the growth of corals will slow down due to the inability to perform photosynthesis and affect oxidase, a condition that causes coral polyps to shrink. According to the inference, this may be due to the initial addition of microplastics to block light sources, causing corals to supplement more light energy and regulate the number of zooxanthellae in the endoderm to produce more zooxanthellae to provide basic nutrients as a short-term resistance. However, after a long period of energy consumption, zooxanthellae can no longer provide sufficient nutrients through photosynthesis, and finally corals start to consume zooxanthellae as an energy source, which eventually causes polyps to shrink (Lanctôt et al., 2020). Therefore, in addition to polluting the ocean, microplastics also pose a threat to marine life.

In the water environment, microplastics will suspend or sink into marine sediments with different densities. Low-density microplastics may also increase in density due to chemical precipitation or biofilm attachment, resulting in sinking to the seabed to mix with sand and gravel, and they may also cover coral reefs. High-density plastics, such as polyvinyl chloride (PVC), sink and accumulate on the seafloor, mixing with seafloor sediments (Browne et al., 2007). Due to their small size and buoyancy, microplastics are easily ingested by aquatic organisms as food or coated on the surface of marine organisms, resulting in stunting their growth. Benthic marine organisms may also ingest microplastics deposited on the seafloor during feeding (Anderson et al., 2016). Numerous studies have demonstrated the existence of microplastics in various marine species, and biological intake of microplastics has become the norm in the entire marine system, which is worrisome. At present, we do not know the threats posed by the long-term accumulation of microplastics in living organisms, which requires long-term monitoring and investigation (Clark et al., 2016).

Although microplastics are inert polymers, they can actually act as a carrier for the adsorption of pollutants, and the chemicals added during the manufacturing process can be toxic to organisms (Pittura et al., 2018). Furthermore, these ingested microplastics may cause harm to organisms higher up the food chain through biomagnification (Fossi et al., 2012).

11.4 WHAT NUTRITION DO MICROPLASTICS HAVE AND WHY DO THEY REPLACE THE STATUS OF LIVE FOOD?

Live food is one of the important foods for marine organisms and their young. During the juvenile or larvae period, marine organisms need to eat live food, and different marine organisms have different mouth sizes, so they need to consume different sizes of live food as a source of nutrition at different growth stages. The live food common to general marine life grows from small to large, such as *Isochrysis galbana* tml, rotifers, *Artemia nauplii*, copepods, or protists (Ding et al., 2021).

The initial live food size of these marine organisms is about 3–500 µm, and marine organisms can ingest different live foods as a source of nutrition at different growth stages. Taking the clown *Amphiprion akallopisos* as an example, *Brachionus plicatilis* can be used as the initial live food for juveniles during reproduction. The size of this species of rotifers is about 162–243 µm (Dhaneesh et al., 2012). However, at different growth stages, in addition to ingesting different sizes of live food, the amount of live food ingested will gradually increase as the juveniles grow (Ding et al., 2021).

According to previous research, microplastics of various sizes are scattered in the ocean, of which plastic fibers are the most commonly found. Because the fibers are washed during the manufacturing process, however, each fiber may be broken down into 1900 fine fibers and released into the environment through wastewater and sludge (Browne et al., 2011). In addition to plastic fibers, the addition of microplastics to everyday products is also a direct source of pollution. Microplastics are often used as exfoliants in facial cleansers and toothpastes. Exfoliators are made from microplastics ranging in size from 70–400 µm, and depending on the amount of microplastic in the exfoliator, up to 94,500 microplastics can be released per exfoliator application (Napper et al., 2015). And the size of this microplastic is exactly the same as the size of live food in the early stage of marine life. Therefore, marine organisms have a high probability of mistakenly swallowing plastic particles when ingesting food, posing a threat to digestion, absorption, growth, and physiology. Organisms living in the ocean at different stages of growth will continue to ingest microplastics, and these microplastics will continue to accumulate in organisms. Current studies have shown that *Mytilus edulis*, *Pomatoschistus microps*, *Epinephelus coioides*, *G. columna*, and other farmed organisms will ingest microplastics and accumulate them in the body (Chen et al., 2022; Devriese et al., 2015). These aquaculture organisms will transfer microplastics to the human body through the food chain to accumulate, which may cause harm to human health.

11.5 THE IMPACT OF MICROPLASTICS ON MARINE ECOSYSTEMS

Over the past 50 years, plastic production has exploded by volume and variety. Such as polyethylene (PE), polypropylene (PP), polyvinyl chloride (PVC), polystyrene (PS), and polyethylene terephthalate (PET) are among the most widely used polymer plastics, but many of them are disposed of after one-time use, and generate excessive plastic waste (Mato et al., 2001). Although these plastic wastes may eventually be incinerated, sent to landfills, or recycled, some may end up in the ocean through sewers, leakage of garbage from landfills, or discharge water from sewage treatment plants. From the perspective of macroscopic observation, it is mainly due to the production of a large amount of plastic waste that it pollutes the ocean. However, according to microscopic examination, because the plastic itself has chemical additives added in the manufacturing process, it accumulates in the sea and then releases persistent pollutants (Mato et al., 2001). Plastic easily adheres to water-based organic pollutants or grease and emits toxic pollutants. Studies have shown that polyethylene microplastic contains a large amount of polychlorinated biphenyls (Endo et al., 2005), and microplastics in the ocean are more difficult to intercept or remove by wastewater treatment equipment than large plastic waste, which will lead to microplastics entering the body of marine organisms through different pathways. The basic consumers of filter feeding in the ocean are not selective in filtering food, so they will not distinguish between live food or microplastic, and microplastics are swallowed directly through filter food, for example, organisms such as mussels, oysters, corals, or sponges. Crabs, shrimps, and so on in benthic organisms may accidentally eat microplastics deposited on the bottom. Other species that forage selectively may ingest microplastics by ingesting contaminated prey.

According to records, microplastics have been found in a variety of plankton. The large number and small size of these organisms belonging to the first stage of the food chain make them more likely to touch and eat microplastics in the ocean (Cole et al., 2013). Plankton is the primary food source for most marine organisms, and microplastics may be transferred to higher-order species due to the relationship of the food chain and then continue to accumulate in higher-order organisms in the food chain, for example, fish, birds, and marine or terrestrial mammals (do Sul et al., 2013). According to statistics, there are at least 170 species of marine vertebrates and invertebrates in the ocean, from benthic to oceanic, that eat microplastics produced by humans by mistake (Vegter et al., 2014). Microplastics are found in marine invertebrates such as sea cucumbers, echinoderms, mussels, lobsters, amphipods, nereis, and barnacles (Graham and Thompson, 2009; Murray and Cowie, 2011). Lusher et al. (2013) found that ten species of fish in the English Channel had up to 36.5% plastic content in their gastrointestinal tracts (Lusher et al., 2013). Through ingestion by organisms and the accumulation of the food chain, microplastics also infiltrate the food chain from the environment and even appear in the seafood that people eat, for example: *Nephrops norvegicus* (Murray and Cowie, 2011). Traces of microplastics are also found in marine mammals. For example, microplastics with a diameter of about 1 mm have been found in the feces of sea lions (McMahon et al., 1999). Creatures in the ocean, from various types of plankton to fish at different growth stages (including larvae and juvenile), have records of ingesting microplastics. It can be seen that the impact of plastic particles on the marine environment is quite extensive. It not only affects the habitat but also threatens the health of marine organisms. Through accumulation in the food chain, microplastics accumulate more in the bodies of predators.

In addition, some studies have pointed out that microplastics have a large overall surface area and adsorption properties, so it is easy for it to adsorb toxic substances in seawater. Plastic raw material particles are lipophilic and have a high affinity for persistent organic pollutants (POPs). Studies have found that the concentration of POPs in microplastics is 1 million times that of the surrounding seawater (Le et al., 2016). In addition, microplastics may release toxic chemicals into seawater, and plasticizers and other chemicals are added during the manufacturing process of plastic products, such as bisphenol A (BPA), nonyl phenol (NP), or polybrominated diphenyl ethers (PBDEs). These substances have been proved to have biological toxicity, such as disturbing physiological metabolism and endocrine function. Once these microplastics containing chemical poisons are ingested by marine organisms, they may be poisoned. Past research has found that exposure of *Melanotaenia fluviatilis* to microplastics contaminated with PBDEs for 21 consecutive days caused the fish to have higher proportions of the relevant chemicals than normal fish populations (Wardrop et al., 2016). It can be seen that chemical substances are mediated by microplastics, and toxins may pass through the water or food chain and gradually accumulate in fish.

Recent studies have pointed out that microplastics can directly affect the physiology, growth, and survival of marine organisms. Due to microplastic ingestion, *Carcinus maenas*, crustaceans, and planktonic copepods, exhibit decreased motility, slower foraging, and reduced energy intake, which in turn affect the reproduction, hatching or egg production (Cole et al., 2013). *Mytilus edulis* is a filter feeder and feeds on microalgae or live food in filter feeding water. Microplastics of 3 or 9.6 μm will remain in the body during the filter feeding process, and microplastics accumulated in the digestive tract for 3 days will be transferred to the circulation system and remain in various organs of the body for up to 48 days or even longer (Browne et al., 2008). *Mytilus edulis* also developed immune and inflammatory responses after ingesting microplastics that did not adsorb chemical pollutants, which means that microplastics themselves can cause cell damage and even death in *M. edulis* (Avio et al., 2015). In order to confirm that the food chain would indeed cause the accumulation of microplastic toxins and cause casualties to predators, *M. edulis* contaminated with microplastic was fed to *C. maenas*, and *C. maenas* also had several microplastics in the body 21 days after eating the contaminated *M. edulis* (Avio et al., 2015). This shows the phenomenon of transfer through the food chain. Plastic raw materials that have not been made into plastic containers have also been found to affect the cranial nerve function of *Pomatoschistus microps* (Oliveira et al., 2013). In addition,

Effect of Microplastics on Marine Ecosystem

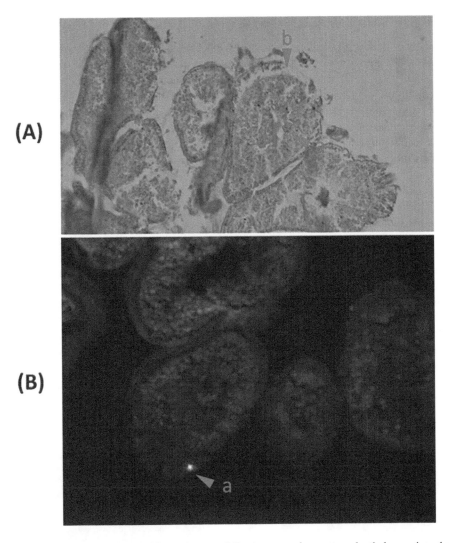

FIGURE 11.3 Tissue damage caused by exposure of *Goniopora columna* to polyethylene microplastic. A: The endoderm of coral contains microplastics and causes mesenteric atrophy and vacuolar degeneration of the basal layer. Arrow a refers to microplastics and vacuolar layer, arrow b refers to ectoderm. B: Microplastics were observed in the endoderm of corals under a fluorescence microscope (×200), and arrow a points to the microplastics.

Oryzias latipes ingested polyethylene microplastic raw materials and also experienced endocrine disruption and liver stress response (Rochman et al., 2013). In addition to fish and shellfish, recent studies have found that corals living on marine reefs also devour microplastic. According to Chen et al. (2022) and Hung et al. (2022) studies, *G. columna* also ingests microplastic, causing damage to endoderm tissue and triggering mesenteric atrophy and vacuolar degeneration, polyp atrophy, mucus hyperplasia, and other physiological phenomena. Microplastic pollution in the ocean shows a serious threat to coral survival (Figure 11.3). It was also found that when *G. columna* was exposed to seawater containing PE microplastic at 5 mg/L, tissue damage occurred after one week, so low concentrations of microplastic could pose a threat to corals (Chen et al., 2022). Therefore, the threat by microplastics to the marine environment and organisms is quite serious, and it is one of the marine pollution problems that we should pay attention to.

11.6 THE THREAT OF MICROPLASTICS TO AQUACULTURE

In the process of marine aquaculture, water source is one of the most important environmental factors. Plastic water pipes or buckets are often used for water diversion. Years of erosion and sunlight may cause the plastic to be weathered or fragmented, forming microplastics deposited in the water supply. In addition, the ocean already contains a lot of microplastics, and these microplastics will be pumped into fishponds along with seawater. In addition to environmental factors, the packaging bags of the feed also contain plastic components, so microplastics may be incorporated into the feed during the feeding process and be swallowed by the fish (Figure 11.4).

Previous studies have pointed out that the hazards of microplastic to aquaculture organisms:

(1) Cytopathy: Exposure of *Mytilus edulis* to seawater containing HDPE > 0–80 μm in size results in proliferation of granulocytoma *in vivo* and reduced lysosomal membrane stability and granulocytoma formation associated with cellular non-neoplastic inflammatory responses. Decreased lysosomal membrane stability is associated with stress and toxic responses and is one of the most sensitive indicators of xenobiotic-induced pathological changes. Therefore, microplastics cause severe immune disease and inflammation in *M. edulis* in a short period of time (Von Moos et al., 2012).

(2) Tissue injury: According to Oliveira et al. (2013), exposure to uncontaminated heavy metals and chemicals of PE microplastic 1–5 μm will result in a decrease in the activity of *Pomatoschistus microps* acetylcholinesterase, regardless of exposure to pyrene-containing microplastic, uncontaminated PE microplastic itself. It can also cause damage to the fish's cranial nerve function. In addition to causing inflammation in digestive and metabolic organs, microplastics also cause obvious inflammation and mucus hyperplasia in the respiratory organs of fish gills. Inflammation can lead to difficulty breathing. In addition

FIGURE 11.4 Grouper feed used in the breeding process; the packaging materials are mostly plastic bags.

to fish, studies have found that microplastics accumulate in the gill tissue of white shrimp, leading to cell inflammation and even cell oxidation or death (Hsieh et al., 2021).

Larger ingested plastic fragments may cause gastrointestinal injury and death after intestinal obstruction. Other harmful effects include impaired secretion of digestive enzymes, decreased feeding rate, steroid hormone imbalance, delayed ovulation and birth, and reproductive failure.

(3) Endocrine disorders: When *Oryzias latipes* ingested uncontaminated microplastic raw materials, endocrine disorders were also found in adult fish, resulting in changes in the endocrine disorder-related genes Vtg I, Chg H, and ERα in female fish and stress responses in the liver, including adipose tissue, vacuolar tissue, and glycogen depletion. Microplastics (40–48 μm) also affect the antioxidant enzymes of *G. columna*, including SOD and CAT, causing a stress response in the physiological metabolism of corals (Chen et al., 2022; Rochman et al., 2013).

11.7 ACTIONS TO PROTECT MARINE ECOLOGICAL SUSTAINABILITY

Microplastics in the ocean are long-term pollution and human-made waste that have been accumulating since they were first developed and used 40 years ago (Thompson et al., 2004, 2005). Sampling from surface water or beach sand has found virgin resin pellets, compounded masterbatch pellets, and smaller fragments of plastics, along with traces of these microplastics (Moore, 2008). These microplastics must be broken down into smaller structures by means of degradation.

Degradation is a chemical change that gradually breaks down the average molecular weight of the polymer. Since the structural integrity of plastics always depends on their molecular weight, any significant degree of degradation will inevitably weaken the material. Most degraded plastics become brittle and break down into powdery fragments when degraded. Even these normally invisible fragments can be further degraded, and the carbon in the polymer is converted into CO_2 and incorporated into marine biomass. When this process is complete and all organic carbon in the polymer has been converted, it is called complete mineralization (Andrady, 2011; Eubeler et al., 2009). Currently known plastic degradation methods in the ocean are shown in Table 11.1.

The current recycling industry technology is still unable to filter and recycle microplastics in the sea through tools. These microplastics will gradually accumulate in the sea and become smaller over time. Currently, microplastic pollution can only be reduced by recycling large plastic waste or reducing the use of plastic products. In addition, recently, decomposable, biodegradable products have been developed for packaging materials, which are mainly made of vegetable fibers or starches (Nazareth et al., 2019). According to a 2019 study, four out of six samples of plastic packaging purchased in Canada, the United States, and Brazil were not degraded after being immersed in seawater for 180 days (Nazareth et al., 2019). Although the industry claims that these environmentally

TABLE 11.1
Degradation Methods of Plastics in the Ocean

Degradation Factor	Mechanism
Biodegradation	Microbial decomposition leads to plastic degradation
Photodegradation	Photooxidative degradation induced by prolonged sunlight exposure (UV-B radiation)
Thermal degradation	High temperature causes plastic to degrade
Thermooxidative degradation	Under the action of medium and high temperature, plastic degradation is gradually carried out
Hydrolysis	Reaction with water

friendly plastics are biodegradable, this may be a technical problem with the product or in order to sell the product as an environmentally friendly product (Nazareth et al., 2019). Therefore, with the current regulation on the use of plastic products around the world, in order to effectively reduce the amount of microplastics in the environment, it is necessary to implement garbage classification and reduce the use of plastic wastes so as to effectively reduce the amount of microplastics in the ocean.

11.8 CONCLUSIONS AND PERSPECTIVES

The ocean is the ultimate gathering place for microplastics. A large amount of microplastics flowing into the ocean will pose a serious threat to the marine ecosystem and marine life. These microplastics travel in the ocean with ocean currents and eventually deposit on the seabed. However, floating microplastics also have a great chance of being swallowed by marine organisms, which will cause chronic poisoning of organisms or affect their survival. In order to reduce the microplastics in the ocean, each country has proposed regulations to restrict the recycling and use of plastic products, but every year, a large number of large plastics drift into the sea and gradually degrade into microplastics. Therefore, effective reduction of plastic waste and good waste classification will reduce the accumulation of microplastics and provide marine organisms with a clean and pollution-free habitat so as to achieve marine ecological sustainability.

REFERENCES

Anderson, J.C., Park, B.J., Palace, V.P., 2016. Microplastics in aquatic environments: implications for Canadian ecosystems. Environmental Pollution 218, 269–280.

Andrady, A.L., 2011. Microplastics in the marine environment. Marine Pollution Bulletin 62, 1596–1605.

Avio, C.G., Gorbi, S., Milan, M., Benedetti, M., Fattorini, D., d'Errico, G., Pauletto, M., Bargelloni, L., Regoli, F., 2015. Pollutants bioavailability and toxicological risk from microplastics to marine mussels. Environmental Pollution 198, 211–222.

Browne, M.A., Crump, P., Niven, S.J., Teuten, E., Tonkin, A., Galloway, T., Thompson, R., 2011. Accumulation of microplastic on shorelines worldwide: sources and sinks. Environmental Science & Technology 45, 9175–9179.

Browne, M.A., Dissanayake, A., Galloway, T.S., Lowe, D.M., Thompson, R.C., 2008. Ingested microscopic plastic translocates to the circulatory system of the mussel, *Mytilus edulis* (L.). Environmental Science & Technology 42, 5026–5031.

Browne, M.A., Galloway, T., Thompson, R., 2007. Microplastic—an emerging contaminant of potential concern? Integrated Environmental Assessment and Management 3, 559–561.

Castro, G.B., Bernegossi, A.C., Pinheiro, F.R., Felipe, M.C., Corbi, J.J., 2020. Effects of polyethylene microplastics on freshwater *Oligochaeta allonais inaequalis* (Stephenson, 1911) under conventional and stressful exposures. Water, Air, & Soil Pollution 231, 1–13.

Chen, Y.-T., Ding, D.-S., Lim, Y.C., Singhania, R.R., Hsieh, S., Chen, C.-W., Hsieh, S.-L., Dong, C.-D., 2022. Impact of polyethylene microplastics on coral *Goniopora columna* causing oxidative stress and histopathology damages. Science of the Total Environment 828, 154234.

Clark, J.R., Cole, M., Lindeque, P.K., Fileman, E., Blackford, J., Lewis, C., Lenton, T.M., Galloway, T.S., 2016. Marine microplastic debris: a targeted plan for understanding and quantifying interactions with marine life. Frontiers in Ecology and the Environment 14, 317–324.

Cole, M., Lindeque, P., Fileman, E., Halsband, C., Goodhead, R., Moger, J., Galloway, T.S., 2013. Microplastic ingestion by zooplankton. Environmental Science & Technology 47, 6646–6655.

Cressey, D., 2016. The plastic ocean. Nature 536, 263–265.

Derraik, J.G., 2002. The pollution of the marine environment by plastic debris: a review. Marine Pollution Bulletin 44, 842–852.

Devriese, L.I., Van der Meulen, M.D., Maes, T., Bekaert, K., Paul-Pont, I., Frère, L., Robbens, J., Vethaak, A.D., 2015. Microplastic contamination in brown shrimp (*Crangon*, Linnaeus 1758) from coastal waters of the Southern North Sea and channel area. Marine Pollution Bulletin 98, 179–187.

Dhaneesh, K., Devi, K.N., Kumar, T.A., Balasubramanian, T., Tissera, K., 2012. Breeding, embryonic development and salinity tolerance of Skunk clownfish *Amphiprion akallopisos*. Journal of King Saud University-Science 24, 201–209.

Ding, D.-S., Sun, W.-T., Pan, C.-H., 2021. Feeding of a scleractinian coral, *Goniopora columna*, on microalgae, yeast, and artificial feed in captivity. Animals 11, 3009.

Ding, J., Jiang, F., Li, J., Wang, Z., Sun, C., Wang, Z., Fu, L., Ding, N.X., He, C., 2019. Microplastics in the coral reef systems from Xisha Islands of South China Sea. Environmental Science & Technology 53, 8036–8046.

do Sul, J.A.I., Costa, M.F., Barletta, M., Cysneiros, F.J.A., 2013. Pelagic microplastics around an archipelago of the equatorial Atlantic. Marine Pollution Bulletin 75, 305–309.

Endo, S., Takizawa, R., Okuda, K., Takada, H., Chiba, K., Kanehiro, H., Ogi, H., Yamashita, R., Date, T., 2005. Concentration of polychlorinated biphenyls (PCBs) in beached resin pellets: variability among individual particles and regional differences. Marine Pollution Bulletin 50, 1103–1114.

Eubeler, J.P., Zok, S., Bernhard, M., Knepper, T.P., 2009. Environmental biodegradation of synthetic polymers I. Test methodologies and procedures. TrAC Trends in Analytical Chemistry 28, 1057–1072.

Fossi, M.C., Panti, C., Guerranti, C., Coppola, D., Giannetti, M., Marsili, L., Minutoli, R., 2012. Are baleen whales exposed to the threat of microplastics? A case study of the Mediterranean fin whale (*Balaenoptera physalus*). Marine Pollution Bulletin 64, 2374–2379.

Gall, S.C., Thompson, R.C., 2015. The impact of debris on marine life. Marine Pollution Bulletin 92, 170–179.

Geyer, R., Jambeck, J.R., Law, K.L., 2017. Production, use, and fate of all plastics ever made. Science Advances 3, e1700782.

Graham, E.R., Thompson, J.T., 2009. Deposit-and suspension-feeding sea cucumbers (Echinodermata) ingest plastic fragments. Journal of Experimental Marine Biology and Ecology 368, 22–29.

Hsieh, S.-L., Wu, Y.-C., Xu, R.-Q., Chen, Y.-T., Chen, C.-W., Singhania, R.R., Dong, C.-D., 2021. Effect of polyethylene microplastics on oxidative stress and histopathology damages in *Litopenaeus vannamei*. Environmental Pollution 288, 117800.

Hung, C.-M., Huang, C.-P., Hsieh, S.-L., Chen, Y.-T., Ding, D.-S., Hsieh, S., Chen, C.-W., Dong, C.-D., 2022. Exposure of *Goniopora columna* to polyethylene microplastics (PE-MPs): effects of PE-MP concentration on extracellular polymeric substances and microbial community. Chemosphere 297, 134113.

Lanctôt, C.M., Bednarz, V.N., Melvin, S., Jacob, H., Oberhaensli, F., Swarzenski, P.W., Ferrier-Pagès, C., Carroll, A.R., Metian, M., 2020. Physiological stress response of the scleractinian coral *Stylophora pistillata* exposed to polyethylene microplastics. Environmental Pollution 263, 114559.

Le, D.Q., Takada, H., Yamashita, R., Mizukawa, K., Hosoda, J., Tuyet, D.A., 2016. Temporal and spatial changes in persistent organic pollutants in Vietnamese coastal waters detected from plastic resin pellets. Marine Pollution Bulletin 109, 320–324.

Lebreton, L., Slat, B., Ferrari, F., Sainte-Rose, B., Aitken, J., Marthouse, R., Hajbane, S., Cunsolo, S., Schwarz, A., Levivier, A., 2018. Evidence that the Great Pacific Garbage Patch is rapidly accumulating plastic. Scientific Reports 8, 1–15.

Lusher, A.L., Mchugh, M., Thompson, R.C., 2013. Occurrence of microplastics in the gastrointestinal tract of pelagic and demersal fish from the English Channel. Marine Pollution Bulletin 67, 94–99.

Mato, Y., Isobe, T., Takada, H., Kanehiro, H., Ohtake, C., Kaminuma, T., 2001. Plastic resin pellets as a transport medium for toxic chemicals in the marine environment. Environmental Science & Technology 35, 318–324.

McMahon, C.R., Holley, D., Robinson, S., 1999. The diet of itinerant male Hooker's sea lions, *Phocarctos hookeri*, at sub-Antarctic Macquarie Island. Wildlife Research 26, 839–846.

Moore, C.J., 2008. Synthetic polymers in the marine environment: a rapidly increasing, long-term threat. Environmental Research 108, 131–139.

Murray, F., Cowie, P.R., 2011. Plastic contamination in the decapod crustacean *Nephrops norvegicus* (Linnaeus, 1758). Marine Pollution Bulletin 62, 1207–1217.

Napper, I.E., Bakir, A., Rowland, S.J., Thompson, R.C., 2015. Characterisation, quantity and sorptive properties of microplastics extracted from cosmetics. Marine Pollution Bulletin 99, 178–185.

Nazareth, M., Marques, M.R., Leite, M.C., Castro, Í.B., 2019. Commercial plastics claiming biodegradable status: is this also accurate for marine environments? Journal of Hazardous Materials 366, 714–722.

Oliveira, M., Ribeiro, A., Hylland, K., Guilhermino, L., 2013. Single and combined effects of microplastics and pyrene on juveniles (0+ group) of the common goby *Pomatoschistus microps* (Teleostei, Gobiidae). Ecological Indicators 34, 641–647.

Pittura, L., Avio, C.G., Giuliani, M.E., d'Errico, G., Keiter, S.H., Cormier, B., Gorbi, S., Regoli, F., 2018. Microplastics as vehicles of environmental PAHs to marine organisms: combined chemical and physical hazards to the Mediterranean mussels, *Mytilus galloprovincialis*. Frontiers in Marine Science 5, 103.

Rochman, C.M., Hoh, E., Kurobe, T., Teh, S.J., 2013. Ingested plastic transfers hazardous chemicals to fish and induces hepatic stress. Scientific Reports 3, 1–7.

Thompson, R., Moore, C., Andrady, A., Gregory, M., Takada, H., Weisberg, S., 2005. New directions in plastic debris. Science 310, 1117.

Thompson, R.C., Olsen, Y., Mitchell, R.P., Davis, A., Rowland, S.J., John, A.W., McGonigle, D., Russell, A.E., 2004. Lost at sea: where is all the plastic? Science 304, 838.

Vegter, A.C., Barletta, M., Beck, C., Borrero, J., Burton, H., Campbell, M.L., Costa, M.F., Eriksen, M., Eriksson, C., Estrades, A., 2014. Global research priorities to mitigate plastic pollution impacts on marine wildlife. Endangered Species Research 25, 225–247.

Von Moos, N., Burkhardt-Holm, P., Köhler, A., 2012. Uptake and effects of microplastics on cells and tissue of the blue mussel *Mytilus edulis* L. after an experimental exposure. Environmental Science & Technology 46, 11327–11335.

Wardrop, P., Shimeta, J., Nugegoda, D., Morrison, P.D., Miranda, A., Tang, M., Clarke, B.O., 2016. Chemical pollutants sorbed to ingested microbeads from personal care products accumulate in fish. Environmental Science & Technology 50, 4037–4044.

Woodall, L.C., Sanchez-Vidal, A., Canals, M., Paterson, G.L., Coppock, R., Sleight, V., Calafat, A., Rogers, A.D., Narayanaswamy, B.E., Thompson, R.C., 2014. The deep sea is a major sink for microplastic debris. Royal Society Open Science 1, 140317.

12 Role of Marine Algae for GHG Reduction/CO$_2$ Sequestration

Strategies and Applications

Xinwei Sun, Andrei Mikhailovich Dregulo,
Su Zhenni, Obulisamy Parthiba Karthikeyan,
Sharareh Harirchi, El-Sayed Salama, Yue Li,
Raveendran Sindhu, Parameswaran Binod,
Ashok Pandey and Mukesh Kumar Awasthi

CONTENTS

12.1 Introduction	157
12.2 Characteristics of Marine Algae	158
12.2.1 Red Algae (Rhodophyta)	158
12.2.2 Brown Algae (Phaeophyta)	159
12.2.3 Green Algae (Chlorophyta)	159
12.2.4 Blue-Green Algae (Cyanobacteria)	160
12.3 Principles and Challenges Associated with Marine Algae Application	160
12.4 Strategies for Promoting Sustainable Marine Algae Applications for GHG Reduction/CO$_2$ Sequestration	160
12.5 Role of Marine Algae in Global CO$_2$ Sequestration: Physiological Mechanisms and Recent Developments for CO$_2$ Sequestration	162
12.6 Research Limitations	163
12.7 Conclusions and Future Perspectives	164
12.7.1 Acknowledgments	165
References	165

12.1 INTRODUCTION

The increasing increase in anthropogenic impact on the atmosphere due to CO$_2$ emissions causes a number of global natural changes. Despite all the efforts of the international community to reduce the anthropogenic impact on the climate, the problem of greenhouse gas emissions is getting worse every year, having an increasing impact on human health. This is especially evident for large world economies. All this involves the search for cost-effective methods to reduce greenhouse gases and CO$_2$, such as photovoltaics (Barati et al., 2022) and bioelectricity. Considering all aspects of the impact of greenhouse gases and CO$_2$ on a global scale, it should be recognized that the main indicator of their impact is the world ocean. Up to 40% of global CO$_2$ is deposited in the world ocean (Cheng et al., 2022). The effect of carbon sequestration by coastal and marine ecosystems (algae and mangrove plants and organisms) has been called "blue carbon" (Cunha et al., 2022). According to some estimates, an increase in CO_2 in the ocean will negatively affect marine ecosystems and human society in the near future (Saxena et al., 2022). Some researchers believe that the erosion of coastal soils of ecosystems with blue carbon in the near future may have the opposite effect—a significant release of CO_2 into the atmosphere—and therefore it is necessary to develop scenarios for managing CO_2 sequestration (Shekh et al., 2022). Perhaps the resources of the ocean are the very object of research that can help humanity in the fight against the threatening growth of CO_2 in the biosphere. We're talking about algae.

12.2 CHARACTERISTICS OF MARINE ALGAE

Currently, more than 30,000 species of algae are known, and it is assumed that the total number of species is more than 1 million (Zeng et al., 2021). Algae are a large, heterogeneous, and polyphyletic group of simple plants that lack roots, stem, and leaves. Algae contain chlorophyll, which is their main photosynthetic pigment. Figure 12.1 shows some species of marine fauna: macro and microalgae and ways of sequestration of CO_2: capture, fixation, deposition. Photosynthesis of algae accounts for up to 50% of the photosynthetic carbon recorded annually by green plants in the earth's biosphere, while photosynthesis of algae increases with increasing N, P, and Fe. Microalgae are sensitive to the light environment, which affects the characteristics of photosynthesis and growth. Diatoms have higher rates of photosynthesis per unit of carbon than dinoflagellate. A study (Zhu et al., 2022) showed that macroalgae (in most Rhodophyta) can be a source of allochthonous carbon due to export to the open ocean and deep ocean areas at a distance of up to 5,000 km from coastal areas.

12.2.1 RED ALGAE (RHODOPHYTA)

Paleological studies indicate that the number of red algae species gradually decreased: an increase in the Early Cretaceous period–early Miocene to 245 species, followed by a decrease to 43 species in the late Pliocene. Red algae have a characteristic red color due to the presence of the phycoerythrin pigment. In some forms, the color is dark red (almost black), in others pinkish. Marine filamentous, leaf-like, bushy, or cortical algae has very complex sexual reproduction. Light conditions directly affect the growth, pigment content, and amount of protein in algae (Zhang et al., 2022). Red algae live mainly in the seas, sometimes at great depths, which is due to the ability of phycoerythrin (a hexamer

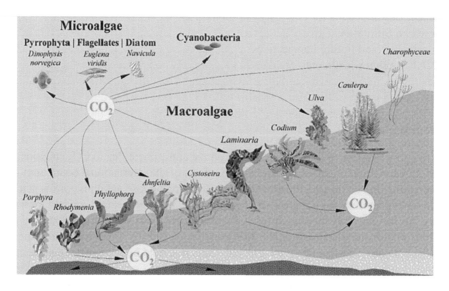

FIGURE 12.1 Typical representatives of marine algae.

of heterodimers consisting of two peptide chains, α- and β-subunits containing 2 and 3 covalently attached chromophores of phycoerythrobilin, respectively) (Zeng et al., 2021) to use green and blue for photosynthesis rays that penetrate deeper than others into the water column. Red algae, unlike brown algae, are able to absorb light penetrating to a depth of up to 25 m. Red algae are considered the most important source of many biologically active metabolites compared to other classes of algae. Agar-agar and other chemicals are extracted from some red algae; for example, porphyra is used for food and is an economically important source of alternative protein (Zahed et al., 2021).

12.2.2 Brown Algae (Phaeophyta)

Brown algae have about 2,000 species, most of which are marine organisms. Individual specimens of brown algae can reach a length of 50 m, for example, giant kelp *Macrocystis pyrifera*. In some brown algae, for example, laminaria, tissue differentiation and the appearance of conductive elements are observed. Multicellular algae layers contain the carotenoid pigment fucoxanthin, which is characterized by a brown color (from olive green to dark brown), which absorbs a large number of rays penetrating to a great depth (Zheng et al., 2022). The efficiency of photosynthesis of algae varies greatly. Photosynthesis of brown algae is carried out at a depth of up to 10 meters. Optimal depths for photosynthesis are up to 20 m; in some cases up to 30 m have been observed in algae in Arctic waters. Some representatives of brown algae emit gas-air bubbles that allow floating forms to hold the stratum on the surface and attached (for example, *Fucus*)—to occupy a vertical position in the water column. The release of gas bubbles during photosynthesis by brown algae *Salicornia gracilaria* is accompanied by a sound from 2 to 20 kilohertz, which may be an indirect indicator of photosynthetic activity of algae (Siddiki et al., 2022). Unlike green algae, many of which grow along the entire length, brown algae have an apical growth point (Zemah-Shamir et al., 2021). Similarly, plasmodesmus, intercellular compounds that play a role in various developmental processes, have been found in brown algae; they are straight channels with a diameter of 10–20 nm, lined with a plasma membrane, and unlike green algae, they lack a desmotubule. Brown algae have a wide range of sexual systems, including both haploid and diploid sexual systems, and in both cases the species can be bisexual or sexually mature (Acién Fernández et al., 2012). Studies (Afonso et al., 2019) show that in brown algae, isogamous species descended from anisogamous ancestors, contrary to the generally accepted model of evolution "from isogamy to anisogamy."

12.2.3 Green Algae (Chlorophyta)

There are 6,000–8,000 species of green algae (most of which are freshwater). Green algae can be both unicellular and multicellular. Green algae cells contain chlorophyll, which gives them the appropriate color, as well as carotene, xanthophyll, and loroxanthin (Ayatollahi et al., 2021) involved in the process of photosynthesis, as well as photoacclematization. In addition to light-absorbing pigments such as chlorophyll and fucoxanthin, green algae have a group of photoprotective carotenoids, which include diatoxanthin, diadinoxanthin, violaxanthin, anthraxanthin, and zeaxanthin, which are involved in the formation of xanthophyll. Green microalgae have unicellular mobile forms (*Chlamydomonas*), unicellular fixed forms (*Chlorella*), multicellular flagellate colonies (*Volvox*), filamentous forms (*Ulothrix, Oedogonium, Spirogyra*), heterotrich forms (*Coleochate*), and siphonal forms (*Acetabularia*) (Almutairi, 2020). Mobile unicellular algae are equipped with flagella. Green algae reproduce asexually, for example, *Trebouxiophyceae* (parts of the stratum, bisection, formation of spores), and sexually, for example, *Chlorophyceae* and *Ulvophyceae*. In some green algae, sexual and asexual reproduction organs may be present on the same specimen; in others, there are sporophytes and gametophytes; for example, *Ulva prolifera* has sexual and obligate asexual populations (Antonaru et al., 2020).

12.2.4 BLUE-GREEN ALGAE (CYANOBACTERIA)

Cyanobacteria (also called blue-green algae) are an ancient group of photosynthetic microbes comprising about 2,000 species in 150 genera, with a wide range of shapes and sizes. Cyanobacteria are gram-negative bacteria. The estimated age of cyanobacteria fossils is 2,000–3,500 million years, which is reason to consider them the parents of the oxygen epoch of the Earth. There is an assumption that photosynthesis spread from procyanobacteria to other types by lateral gene transfer. There are three main ecological groups of cyanobacteria: low-temperature, low-temperature copiotrophs, and high-temperature oligotrophs. The fundamental difference between algae and cyanobacteria is that the latter are a group of prokaryotic bacteria, and algae are small eukaryotic plant organisms (Carone et al., 2022).

12.3 PRINCIPLES AND CHALLENGES ASSOCIATED WITH MARINE ALGAE APPLICATION

The absorption of CO_2 and other greenhouse gases by algae opens up the potential for a wide range of value-added products. Around the world, people use about 291 species of seaweed, mainly for the production of food and hydrocolloids (for example, alginates, agar, and carrageen (Dammak et al., 2022), as well as for the manufacture of medicines, paper, fertilizers, and animal feed. The production of polysaccharides from algae is up to 86,100 tons per year. However, it is important to consider the industrial potential of prokaryotic and eukaryotic algae species. For example, only eukaryotic microalgae are capable of naturally producing triacyl-glycerides, unlike prokaryotic microalgae (cyanobacteria) and macroalgae. Nevertheless, the potential of algae products is incredibly high. Carbon captured and transformed by algae into biomass can be used as a source for the production of bioenergy and other value-added products (Paul et al., 2021).

The commercial potential of microalgae lies in relatively minimal costs for metabolic processes. A prerequisite for the creation of industrial microalgae factories is the creation of an effective system of genetic transformation and editing of algae genes/genomes, for example, using CRISPR-Cas9 (Carone et al., 2022), and transcription activator-like effector nucleases (TALENs) (Dammak et al., 2022). A promising direction for obtaining value-added products is the study of biostimulating effects of microalgae extracts. The joint cultivation and application of microalgae extracts with biofertilizers can improve the quality and quantity of agricultural products (Antonaru et al., 2020). The biostimulating effect of bioextracts obtained by *Aphanothece* sp. and *Chlorella ellipsoidea* affects the growth of tomato roots and shoots ~112% ~53.7%, respectively, which was accompanied by an increase in absorption of N~185%, P~119%, and K~78% in terms of dry biomass (Muthukrishnan, 2022). Along with the search for increasing the volume and quality of agricultural products, the search for an alternative protein for nutrition is also the most important direction in the development of the food algae industry (Carone et al., 2022), as well as the production of bioactive algae peptides used in the pharmaceutical, food, and cosmetic industries. The sorting properties of preparations obtained from algae are effectively used for methods of extraction and control of metals in surface waters and for wastewater treatment systems in photobioreactors (Benner), further optimizing the technology of obtaining oil from algae (belt presses are the recommended technology for dehydrating algae before oil extraction after wastewater treatment). Thus, with obvious prospects for the use of algae on an industrial scale for a number of industries, it still faces 1) the profitability of the production of preparations derived from algae, and 2) the production of heat from microalgae is accompanied by higher greenhouse gas (GHG) emissions than emissions from fossil fuels.

12.4 STRATEGIES FOR PROMOTING SUSTAINABLE MARINE ALGAE APPLICATIONS FOR GHG REDUCTION/CO_2 SEQUESTRATION

The cultivation of microalgae for CO_2 sequestration and biomass production was well studied using different types of photobioreactors or onshore cultivation platforms, but fewer efforts were focused on offshore cultivations. Offshore cultivations of macro- and microalgae reduce or eliminate the demands for fresh water, fertilizers, and cultivation land. Large algal farms for human and animal consumption are commonplace in Asia, Oceania, and parts of northern Europe. Considering the food-fuel-water energy nexus and pharmaceutical applications, the cultivation of algae using sea water is considered, mainly focused in recent years on the blue bioeconomy. As of 2016, the global production of algal biomass (fresh weight) was reported to be 32.67 Mt (Nishshanka et al., 2022).

In the United States, 5 billion gallons of algal biofuel production was targeted by 2030. Areas of the South Atlantic and the Gulf of Mexico, as well as the West Coast, Alaska, Hawaii, and other Pacific Islands, have been identified as preferred geographic regions for macroalgal biomass production, with portions of Hawaii, California, Arizona, New Mexico, Texas, Louisiana, Georgia, and Florida as potential areas with adequate sunlight for optimal open cultivation of microalgal biomass within the United States. Additionally, areas of the southwestern United States have been identified as the most suitable for closed systems for growing algae, such as photobioreactors. There is several finding and projects were developed and tested to prove the significant impact of algae for CO_2 sequestration and carbon capture. Specifically, the Offshore Membrane Enclosures for Growing Algae (OMEGA) project was proposed that uses PBRs to cultivate freshwater algae using wastewater as nutrient sources. The OMEGA project was located in San Francisco and can treat the wastewater discharged from the San Francisco Southeast Wastewater treatment plant with a total capacity of $1.23 \times 10^6 m^3$ with an extension of 11. The photobioreactor tubes were made of clear linear low density polyethylene connected with cam-lock fittings to a U-shaped PVC manifold. The PBRs had six swirl vanes that move on a helical path to mix the algae during cultivation and improve photosynthesis. The PBRs were fitted with all the required sensors to measure the pH, ORP, DO, temperature, and so on, and data were recorded and continuously monitored from control stations offshore.

A mixed culture was used in PBRs; however, it was dominated by *Desmodesmus* sp., with ammonium removal of >90% from the wastewater and 50% higher CO_2 fixation capacities. The algal productivity ranged between 14.1 and 20 $g/m^2/day$, which was higher than the average productivity of 13.2 $g/m^2/day$ for US production platforms. The CO_2 fixation rate of *Desmodesmus* sp. is reported to be 0.26 $gL^{-1}day^{-1}$. In a different study by Algal System LLC at Daphne, Alabama, the dominant cultures from offshore PBRs treating wastewater were enriched with the genera *Scenedesmus*, *Chlorella*, *Cryptomonas*, *Micractinium*, *Desmodesmus*, *Chlamydomonas*, *Euglena*, *Pandorina*, *Coelastrum*, and filamentous cyanobacteria (*Geitlerinema*). Small coccoid algae (<5 μm) were also counted but were not identified. The floating PBRs were able to produce 3.5–22.7 $g/m^2/day$ of biomass during continuous operations, while production was affected by the prevailing temperature and the harvesting frequency. Several marine macro- and microalgae cultivations are possible offshore. The several proposed semi-porous container for the algal cultivation nearshore are in the early stages of development. But algae such as seaweed are most commonly cultivated using offshore ropes on the sea bed, and harvesting is easy to compared to that of algal cultivation systems. Cultivation area siting should be in less energetic areas such as slower currents, and reduced sea states are very important. Also, as mentioned earlier, temperature and harvesting technologies are critical. Previous studied has already identify the effects of winds, waves, and water currents on the hydrodynamic structures of the floating PBRs on offshore establishments. Alishah et al. (2019) developed a multiscale model for *Ulva* sp. for intensive cultivation farming that accounts for growth and nutrient consumption as key parameters. Semi-permeable membranes are generally used in floating PBRs for improved nutrient transformation for biomass growth. These membranes are usually operated as forward osmosis principles to remove the biomass as one of the harvesting techniques.

Also, different types of photobioreactors were developed and tested for offshore cultivation of microalgae, including i) a combination of raceway pond and airlifts reactors, ii) tubular module PBRs (Figure 12.2a); and iii) a simple floating pet bottles (Figure 12.2b). Designing PBRs or cultivation platforms for the mass cultivation of algae offshore should consider critical parameters such as (i) light penetration and culture depths in the reactors, (ii) turbulence for better gas and nutrient

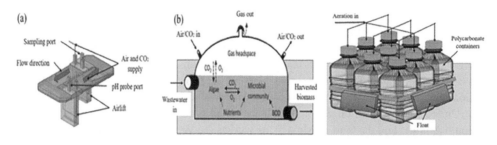

FIGURE 12.2 Different reactor configurations for offshore cultivation of algae.

mixtures, (iii) growth conditions and patterns of selected strains or mixed cultures, and (iv) CO_2 assimilation and nutrient demands. Other than operational parameters and reactor requirements, environmental factors need to be taken into consideration while developing offshore cultivation, such as hydrodynamic loads, local climatic conditions, and long-term environmental variables (such as temperature, water level, CO_2 absorption capacity, high, and low-tides). The temperature increase will significantly affect solubility as well as photosynthetic fixation capacities. Specifically, atmospheric CO_2 will be less sufficient for biomass growth; therefore, additional CO_2 sources should be considered. Therefore, offshore cultivation of algae is a feasible and reliable technology to reduce the atmospheric concentrations of CO_2 and reduce the global warming potential. As of today, only a very few projects and prototypes have been tested for algae cultivation, and further studies are required to better understand the techno-economics of offshore cultivations.

12.5 ROLE OF MARINE ALGAE IN GLOBAL CO_2 SEQUESTRATION: PHYSIOLOGICAL MECHANISMS AND RECENT DEVELOPMENTS FOR CO_2 SEQUESTRATION

Typically, plants and oceans play a critical role in the mitigation and sequestration of CO_2; however, this natural mechanism can reduce 50% of CO_2, and the leftover CO_2 increasingly collects in the Earth's atmosphere. Recently, scientists have focused on algal biomass for CO_2 sequestration to reduce greenhouse gas emissions and their side effects, especially global warming. Algae have higher growth rates rather than plants (ca. 100 times) and do not need fertile lands, which leads to considerable capacities for novel CO_2 sequestration approaches (Alishah Aratboni et al., 2019). Additionally, algal products such as bioactive materials, cosmetic components, alginate, carrageenan, agar, β-carotene, astaxanthin, animal feedstocks, carbohydrates, proteins, dietary supplements, or biofuels (biodiesel, bioethanol, or biohydrogen) that are produced during CO_2 sequestration may result in a cost reduction of the process (Alba and Kontogiorgos, 2019). For example, in a study, mathematical modeling on a 200 MW natural gas combined cycle plant and *D. salina* (as CO_2 fixator and carotenoids producer) showed that the price for the CO_2 sequestration would be US$146.9/tons CO_2 (Afonso et al., 2019). In general, algae form a large and diverse photosynthetic group of eukaryotic organisms and microorganisms, which are widespread in marine and terrestrial environments (Barati et al., 2022). Red, green, and brown algae, and even diatoms, are considered micro- and macroalgae. Among them, brown algae (phaeophytes) live in marine environments, and red algae (rhodophytes) are mostly marine algae, while green algae (chlorophytes) live in freshwater, and only a few species such as *Dunaliella*, *Pyramimonas*, and *Tetraselmis* are classified as marine microalgae. These micro- and macroalgae have considerable potential to be used in approaches that capture CO_2 to reduce its emission and consequent global warming. Algal cells can capture CO_2 through photosynthesis, which has two main stages (light-dependent and -independent stages). During the light-dependent stage, energy-carrying molecules store sunlight energy. In the light-independent stage, the CO_2 is entrapped and converted into organic compounds through the

Calvin-Benson-Bassham cycle with the aid of charged energy-carrying molecules. Theoretically, microalgae can annually convert approximately 9% of solar energy and consume 530 tons of CO_2 to produce dry biomass (280 tons/ha); however, commercial production needs more improvement to approach the theoretical yield.

Marine and freshwater algae can grow in open or closed systems, ponds, or photobioreactors to capture CO_2 from various sources such as concentrated CO_2 streams; flue gas of power stations; and other production units such as aluminum, iron, steel, cement, mining, petrochemicals, fertilizers, and pulp and paper industries (Paul et al., 2021). Open pond systems are particularly used for large-scale cultivation of microalgae, including raceway ponds and multi-layer bioreactors. Up to now, various types of PBRs such as airlift, tubular, flat plate, sequence batch membrane carbonation, internally illuminated, bag, membrane, and filtration have been developed as closed cultivation systems to facilitate microalgal CO_2 sequestration (Shekh et al., 2022). Both marine and freshwater microalgae can capture CO_2; however, based on location; impurity of gaseous streams (such as SO_x and NO_x); gas concentration; and other factors such as light irradiation, uniformity of mixing, cell density, oxygen accumulation, temperature, pH, salinity, or biomass production, appropriate species and running systems should be selected (Barati et al., 2022). For example, the marine cyanobacterium (formerly blue-green alga) *S. elongatus* can sequester CO_2 derived from flue gas at high temperatures up to 60°C, which is an advantage for the treatment of flue gas with high temperatures. On the other hand, at high temperatures, oxygen solubility and its ratio to CO_2 increase, which significantly affects the oxygenase activity of ribulose 1, 5-bisphosphate carboxylase/oxygenase (RuBisCO) and decreases its affinity for CO_2, resulting in a lower yield of CO_2 sequestration. The tolerance or resistance of algal strains to process conditions of CO_2 sequestration is an important factor that affects the process's productivity and yield. For instance, *Dunaliella* is an algal genus that is highly fragile and sensitive to shear stress (due to the lack of cell walls), while the genus *Arthrospira* can easily tolerate this stress. Another significant factor that influences the efficiency of CO_2 fixation by algal cells is the risk of contamination. Algal species that are more resistant to microbial contamination are preferred for large-scale CO_2 sequestration.

One of the novel and sustainable approaches in which inorganic and organic carbon are converted to algal biomass is phytoremediation. This method can be used for various wastewater treatments, as high amounts of carbon, nitrogen, phosphorus, and even heavy metals exist (Cheng et al., 2022). Phytoremediation is based on three metabolic pathways of algal cells, heterotrophic, photoautotrophic, and mixotrophic metabolisms (Siddiki et al., 2022). In heterotrophic growth, organic compounds are assimilated by the algal cell under aerobic respiration, while under mixotrophic conditions, inorganic and inorganic carbon are simultaneously utilized under irradiation and dark conditions, respectively. In this approach, regarding resource recovery of wastewater and sustainability of the process, the choice of algal cells depends on lipid accumulation and cell metabolism capacities (Cunha et al., 2022).

12.6 RESEARCH LIMITATIONS

Algae can reproduce, tolerate extreme conditions, and have high photosynthetic efficiency (Cheng et al., 2022). The use of algae for carbon sequestration has great potential for industrialization (Afonso et al., 2019). The use of marine algae for large-scale carbon sequestration economies has so far been infeasible (Almutairi, 2020). To achieve economic viability and sustainability, major barriers in both upstream and downstream processes must be overcome. The upstream process depends on the selection of algae strains, nutrient availability and culture conditions, and environment. However, downstream processing is limited by the method of harvesting, cell crushing, and extraction (Barati et al., 2022). The use of CO_2 from flue gas by marine microalgae to reduce carbon emissions to the atmosphere is a challenging idea, but several important drawbacks need to be addressed first: (i) at high CO_2 concentrations (>5%), the growth of marine algae may be inhibited. Considering that CO_2 concentrations in flue gases range from 3–30%, it is a critical task to identify marine algae strains that are adapted to grow at high CO_2 concentrations. (ii) Another challenge is the inhibitory effect of sulfur oxides (SO_x), nitrogen oxides (NO_x), and heavy metals on the growth of algae. Small amounts of NO_x and SO_x can act as nutrients for marine algae. However, the amounts of these compounds in flue gases can be toxic to algal cells. The response of marine algae to compounds in flue gas varies from species to species. (iii) Another concern with the use of CO_2 from power plants for algae culture is that the gas is discharged into the stack at too high a temperature (65–95°C in most cases). Consequently, before application to culture, the flue gases must pass through a cooling system to reduce the temperature to an optimum value suitable for algae growth (Cheng et al., 2022). It is also necessary to consider the non-essential cost losses caused by the escape of harmful gases into the environment. This concern can be solved through the collaboration of interdisciplinary scientists in the fields of chemistry, biotechnology, materials, manufacturing, and engineering (Dammak et al., 2022). Therefore, each stage must be closely monitored to prevent any unfortunate events that may have a profound impact on the rest of the chain. Developing environmentally sustainable, green economy practices is essential to creating a carbon-neutral future. If properly designed and operated, marine algae biorefinery production can make a positive contribution to the progress of carbon fixation.

12.7 CONCLUSIONS AND FUTURE PERSPECTIVES

Marine algae have higher CO_2 tolerance, growth rate, photosynthetic activity, and carbon assimilation efficiency than other terrestrial plants. Research on CO_2 sequestration in marine algae has also focused on improving RuBisCO, pyrenoids, and photosynthesis. A few recently developed strategies associated with synthetic biology and other bioengineering techniques have been used to improve CO_2 fixation by redesigning metabolic pathways in the algal cell. The advances in biorefineries for the production of biofuel and value-added products also have the potential to achieve large-scale carbon sinks. Thus, the economical and sustainable implementation of large-scale carbon sequestration by marine algae depends on the improvement of factors (such as growth rate, the understanding of photosynthetic mechanisms, the construction of technological infrastructure, and carbon cost considerations). Future outlooks to achieve better performance of marine algae for CO_2 sequestration are itemized in the following:

(1) Development of low-cost, high-performance PBRs is currently underway. The density of marine algae cannot be too high due to the limitations of light intensity and light transfer characteristics. The exhaust gas transport and uptake need to be considered, so both raceway ponds and flat PBR_S need to be optimized.
(2) Marine algae production is a zero-emission process and requires low energy for cultivation on a large scale. Marine algae energy conversion technology integrated with other biorefineries will contribute to carbon sequestration and biofuel generation.
(3) Several algae growth stages have different CO_2 requirements. The conversion from gaseous to liquid state during carbon fixation involves thermodynamic and kinetic energy. Therefore, the cultivation of algae should be studied for the process parameters and external factors.
(4) The mechanisms of energy conversion in marine algal carbon sequestration studies have not been fully investigated, and the main pathways involved in free energy depletion have not been quantified. This is due to the lack of a proper macro-level energy conversion model and theoretical basis to explain the phenomenon. Therefore, the adaptability of algal growth and the operability of the corresponding optical systems should be investigated through cross-disciplinary approaches.
(5) Most of the screening concerning marine algae still uses model algae. Therefore, metabolite isolation, growth physiology, resistance to external factors, and genetic flexibility of the strain also need to be considered. Genomics and transcriptomics are advanced

tactics to provide intrinsic information on algal biology, and it is beneficial to enhance the productivity of marine algae.
(6) The combination of synthetic biology with other biotechnologies to control gene expression and manipulate metabolic pathways (using an effective toolkit [CRISPR/CAS9] and re-engineering optimized metabolic pathways) can help marine algae to develop photosynthetic efficiency for large-scale CO_2 sequestration.

12.7.1 Acknowledgments

The authors are grateful for the financial support from the Shaanxi Introduced Talent Research Funding (A279021901) and the Introduction of Talent Research Start-up fund (Z101022001), College of Natural Resources and Environment, Northwest A&F University, Yangling, Shaanxi Province 712100, China. We are also thankful to all our laboratory colleagues and research staff members for their constructive advice and help.

REFERENCES

Acién Fernández FG, González-López C, Fernández Sevilla J, Molina Grima E. Conversion of CO_2 into biomass by microalgae: how realistic a contribution may it be to significant CO_2 removal? Applied Microbiology and Biotechnology 2012;96(3):577–586.

Afonso NC, Catarino MD, Silva AMS, Cardoso SM. Brown macroalgae as valuable food ingredients. Antioxidants (Basel) 2019;8(9):365.

Alba K, Kontogiorgos V. Seaweed polysaccharides (agar, alginate carrageenan). Encyclopedia of Food Chemistry 2019:240–250.

Alishah Aratboni H, Rafiei N, GaRcia-Granados R, Alemzadeh A, Morones-Ramírez JR. Biomass and lipid induction strategies in microalgae for biofuel production and other applications. Microbial Cell Factories 2019;18:178.

Almutairi AW. Effects of nitrogen and phosphorus limitations on fatty acid methyl esters and fuel properties of *Dunaliella salina*. Environmental Science and Pollution Research 2020;27(26):32296–32303.

Antonaru LA, Cardona T, Larkum AWD, Nürnberg DJ. Global distribution of a chlorophyll of cyanobacterial marker. The ISME Journal 2020;14:2275–2287.

Ayatollahi SZ, Esmaeilzadeh F, Mowla D. Integrated CO2 capture, nutrients removal and biodiesel production using *Chlorella vulgaris*. Journal of Environmental Chemical Engineering 2021;9(2):104763.

Barati B, Zafar FF, Qian L, Wang S, Abomohra AEF. Bioenergy characteristics of microalgae under elevated carbon dioxide. Fuel 2022;321:123958.

Carone M, Alpe D, Costantino V, Derossi C, Occhipinti A, Zanetti M, et al. Design and characterization of a new pressurized flat panel photobioreactor for microalgae cultivation and CO2 bio-fixation. Chemosphere 2022;307:135755.

Cheng P, Li Y, Wang C, Guo J, Zhou C, Zhang R, et al. Integrated marine microalgae biorefineries for improved bioactive compounds: a review. Science of the Total Environment 2022;817:152895.

Cunha SA, Pintado ME. Bioactive peptides derived from marine sources: biological and functional properties. Trends in Food Science & Technology 2022;119:348–370.

Dammak M, Ben Hlima H, Tounsi L, Michaud P, FendrI I, Abdelkafi S. Effect of heavy metals mixture on the growth and physiology of *Tetraselmis* sp.: applications to lipid production and bioremediation. Bioresource Technology 2022;360:127584.

Muthukrishnan L. Bio-engineering of microalgae: challenges and future prospects toward industrial and environmental applications. Journal of Basic Microbiology 2022;62(3–4):310–329.

Nishshanka GKSH, Anthonio RADP, Nimarshana PHV, Ariyadasa TU, Chang JS. Marine microalgae as sustainable feedstock for multi-product biorefineries. Biochemical Engineering Journal 2022;108593.

Paul S, Bera S, Dasgupta R, Mondal S, Roy S. Review on the recent structural advances in open and closed systems for carbon capture through algae. Energy Nexus 2021;4:100032.

Saxena A, Mishra B, Sindhu R, Binod P, Tiwari A. Nutrient acclimation in benthic diatoms with adaptive laboratory evolution. Bioresource Technology 2022;351:126955.

Shekh A, Sharma A, Schenk PM, Kumar G, Mudliar S. Microalgae cultivation: photobioreactors, CO2 utilization, and value-added products of industrial importance. Journal of Chemical Technology & Biotechnology 2022;97(5):1064–1085.

Siddiki SYA, Mofijur M, Kumar PS, Ahmed SF, InAyat A, Kusumo F, et al. Microalgae biomass as a sustainable source for biofuel, biochemical and biobased value-added products: an integrated biorefinery concept. Fuel 2022;307:121782.

Zahed MA, Movahed E, KhodAyari A, Zanganeh S, Badamaki M. Biotechnology for carbon capture and fixation: critical review and future directions. Journal of Environmental Management 2021;293:112830.

Zemah-Shamir S, Zemah-Shamir Z, Tchetchik A, Haim A, Tchernov D, Israel Á. Cultivating marine macroalgae in CO_2-enriched seawater: a bio-economic approach. Aquaculture 2021;544:737042.

Zeng J, Wang Z, Chen G. Biological characteristics of energy conversion in carbon fixation by microalgae. Renewable and Sustainable Energy Reviews 2021;152:111661.

Zhang C, Dong H, Geng Y, Song X, Zhang T, Zhuang M. Carbon neutrality prediction of municipal solid waste treatment sector under the shared socioeconomic pathways. Resources, Conservation and Recycling 2022;186:106528.

Zheng H, Ge F, Song K, Yang Z, Li J, Yan F, et al. Docosahexaenoic acid production of the marine microalga *Isochrysis galbana* cultivated on renewable substrates from food processing waste under CO_2 enrichment. Science of the Total Environment 2022;848:157654.

Zhu X, Lei C, Qi J, Zhen G, Lu X, Xu S, et al. The role of microbiome in carbon sequestration and environment security during wastewater treatment. Science of the Total Environment 2022;837:155793.

Index

A

adsorbents, 126
agarans, 34
agarose, 34
algal pigments, 2, 3
    biological safety, commercial scope, 12
    de novo synthesis, 3
alginates, 38
amino acids, 93
anti-angiogenic activity, 57
antibacterial action, 83
anticancer activity, 35, 55
anti-diabetic action, 83
anti-inflammatory activity, 55
antiobesity, 35
anti-obesogenic, 83
antioxidants, 64
    activity, 35, 57, 83
aquaculture, 61, 121
aquafeed, 90, 93
    agar, 34
    composition, 62
    functional feed, 62
    plant-based, 90
    reduced, 93
aquatic biomass, 21
Arabinan, 36
Arabinogalactans, 36
astaxanthin, 4, 6, 12, 137
    bioactivity, 5

B

banana, 96
    peel, 96
β-carotene, 11, 13
    bioactivity, 11
bioactive polysaccharides, 32
biochar, 121, 122
    characteristics, 124
    chemical properties, 125
    production, 122
biochemical composition, 52
blue-green algae, 160
brown seaweeds, 38, 159

C

Cacao, 96
    pod husks, 96
Calvin-Benson-Bassam cycle, 163
canthaxanthin, 14
capacitive deionization, 128
carbohydrate, 52, 63
carotene, 11
carotenoids, 1, 4, 12, 15, 54

market, 16
production, 13
    downstream processing, 15
    factors affecting, 13
    outdoor environment, 14
    upstream, 13
structure, classification, significance, 4
carrageenans, 33
catalyst, 126
catalytic decomposition, 121
chitin, 136, 138
chitosan, 137, 139
Chlorella ellipsoidea, 9
Chlorella pyrenoidosa, 13
Chlorella sorokiniana, 6, 7, 8
Chlorella vulgaris, 55
chlorophyll, 2, 15, 46
Chlorophyta, 21, 159
cholesterol, 91
Chromochloris zofingiensis, 5, 6, 9, 11
Citrus lemon, 97
CO_2 sequestration, 160
collagen, 78, 81, 135, 138
    marine, 78
        applications, 79
        biological action, 83
        extraction, 78
        properties, 81
cosmetic application, 84
C. protothecoides, 15
cyanobacteria, 50, 160
cytopathy, 152

D

Dunaliella salina, 11
Dunaliella sp, 46

E

Eichhornia crassipes, 96
Enzymatic extraction, 28, 140
Eukaryotic microalgae, 50, 52

F

feed conversion ratio (FCR), 109
fenton-like oxidation, 121
filler feed, 61
    ingredient, 63, 67
fillers, 62
flash carbonization, 123
folate, 95
food supplement, 55
fucoidans, 38
funorans, 34

G

galactans, 36
gelatin, 135, 138
geographic information system, 115
GHG reduction, 160
green algae, 159

H

Haematococcus pluvialis, 5, 6, 46, 55
histidine, 81
hydrochar, 124
hydrostatic pressure, 140
hydrothermal, 27, 124
    carbonization, 124
    processing, 27
hydroxyapatite, 136, 139

I

immunomodulatory, 34
immunostimulants, 95
inositol, 95

L

laminarans, 38
lemon peel, 97
lipids, 53, 64, 93, 136
    peroxidation, 8
lutein, 6, 7
    bioactivities, 8

M

mannans, 37
marigold flowers, 9
marine
    algae, 1, 158, 160
    collagen, 78
        biological action, 83
        extraction, 78
        properties, 81
    microalgae, 51, 56, 83
        applications, 56
        composition, 51
        filler feed, 63
membrane separation technology, 141
memory function, 9
mevalonate pathways, 3
microplastics, 145, 152
    impact, 149
microwave, 23
mixotrophy, 2, 5
Musa acuminate, 96

N

neuroprotective activity, 57
nucleotides, 92

O

obesity, 10
organic, 92, 121
    acids, 92
    pollutants, 121
oxidative radicles, 12

P

pantothenic acid, 95
peroxisome receptors, 6
pesticides, 14
Phaeophyta, 21, 159
photosynthetic bacteria, 50
phycobilin, 2
phytosterols, 54
pigments, 64
    microalgae, 12
polyphenols, 54
polysaccharides, 31, 33, 34
    bioactive, 32
    biological properties, 31
    extraction, 32
    physiological functions, 31
porosity, 83
porphyrans, 34
prawn, 103
    production, 9
protein, 52, 134
protein hydrolysates, 134
pulsed electric field, 25, 140
    non-thermal technology, 25
pyrolysis, 122

R

reactive extrusion, 28
red algae, 32, 158
    extraction, 32
        polysaccharides, 32
red algae, 32, 158
    structure, 33
regenerative medicine, 84
Rhodophyta, 21

S

scaffolds, 84
Scenedusmus incrassatulus, 14
S. dimorphus, 15
seafood, 131
    discards, 131
    by-products, 131
        environmental impact, 133
        source, 132
    seaweed, 22
        brown, 38
        green, 36
        polysaccharide, 34, 36, 38, 96
        red, 32

Index

selenium, 92
shrimp, 103–105, 107, 108
   culture, 104
   disease control, 114
   farming, 105
      sustainability, 106
   feed, 108
   seed, 107
shrimp farming, 104, 107, 109
singlet oxygen, 12
sources, 145
spirulina, 46
subcritical water extraction, 141
sulfated polysaccharides, 33
supercritical fluid extraction, 141
supercritical fluids, 26

T

taurine, 90
tensile strength, 82
Theobroma cacao, 96
torrefaction, 123

U

ultrasound, 24
   equipment, 24
   extraction, 141
ulvans, 36, 37

V

vitamins, 53, 93

W

water hyacinth, 96
water remediation, 126
wound healing, 85

X

xanthophyll, 4
xylans, 37

Z

zeaxanthin, 9
   bioactivities, 9
   extraction, 10

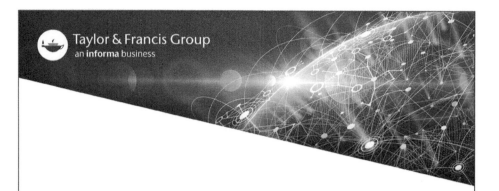